ORCHIDS
OF SOUTH AFRICA | A Field Guide

Steve Johnson
Benny Bytebier

Photography by
Herbert Stärker

D1705925

Published by Struik Nature
(an imprint of Penguin Random House South Africa (Pty) Ltd)
Reg. No. 1953/000441/07

The Estuaries No 4, Oxbow Crescent, Century Avenue, Century City, 7441
PO Box 1144, Cape Town, 8000 South Africa

Visit **www.randomstruik.co.za**
and join the Struik Nature Club for updates,
news, events and special offers.

First published in 2015 by Struik Nature
1 3 5 7 9 10 8 6 4 2

Publisher: Pippa Parker
Managing editor: Helen de Villiers
Project manager: Sarah Gruft
Editor: Emsie du Plessis
Design director: Janice Evans
Design: Tessa Fortuin & Neil Bester
Proofreader: Tina Mössmer

Reproduction by Hirt & Carter Cape (Pty) Ltd
Printed and bound by Paarl Media, Jan van Riebeeck Avenue,
Paarl, South Africa.

ISBN: 978 1 77584 139 5
E-pub: 978 1 77584 263 7
PDF: 978 1 77584 262 0

Photographic credits: Photographs supplied by Herbert Stärker, with further contributions by:
Uli Ade, Jill Blignaut, Benny Bytebier, Gareth Chittenden, Ruth Cozien, Hildegard Crous,
Lourens Grobler, Karin Hansen, Steve Johnson, Judd Kirkel, Patrick Lane, Bill Liltved, Peter Linder,
Mervin Lotter, Douglas McMurtry, Geoff Nichols, Greg Nicolson, Marc Nicolson, Ted Oliver,
Martin Rautenbach, Peter Swart, Ross Turner, Timo van der Niet, Jan Vlok, Louis Vogelpoel,
Martin von Fintel, Karsten Wodrich, Peter Wragg, Bart Wursten and Helmuth Zelesny.

Cover image: *Ceratandra grandiflora,* Joubertina 15.12.2010
Title page Image: *Bartholina burmanniana*

CONTENTS

PREFACE

It has been more than 30 years since the last publication of a field guide covering the entire South African orchid flora. *Wild Orchids of Southern Africa,* published in 1982, soon went out of print and is now a collector's item. *Orchids of Southern Africa,* a reference work published in 1999, remains indispensable to those who have a copy, but only a few hundred copies remain as the publisher lost most of the stock in an accident soon after the book was launched. *Field Guide to the Orchids of Northern South Africa and Swaziland,* published in 2008, filled some of this gap for the north of the country, whereas *The Cape Orchids: A Regional Monograph of the Orchids of the Cape Floristic Region,* published in 2012, did the same for the Cape region (though with a combined weight of 7kg, the two volumes of the latter hardly constitute a suitable field guide!). One of the major challenges for producing a field guide to South African orchids is that several species are highly elusive, some appearing only every 20–30 years after a fire, and thus difficult to capture in photographic images. This was reflected in the fact that many of the species included in the original *Wild Orchids of Southern Africa* were not represented by photographs or, if they were, only by very poor ones that did not allow for definitive identification.

The situation changed in 2006 when Austrian photographer Herbert Stärker began to make regular visits to South Africa as part of a focused mission to produce high-quality photographic images of orchids in the region. We became acquainted with Herbert through his extraordinary photographic contributions (along with those of Bill Liltved)

to *The Cape Orchids* volumes. Thus when Herbert suggested the use of his images for a comprehensive field guide to South African orchids, we readily agreed that the time was right for such a project. Herbert and his wife Helga had also acquired a tremendous amount of new knowledge regarding the distribution, habitat and flowering times of South African orchids during the many years they had spent photographing species in the field.

Orchids of South Africa: A Field Guide is intended for those individuals who enjoy finding and identifying South African orchids in their natural habitat and who want to learn something about their ecology, conservation status and relationship to other orchid species. Of the 473 orchid species known to occur in South Africa, Lesotho and Swaziland, 455 are illustrated in this book. The 18 species that are not featured are very poorly known; approximately half of these have not been seen for more than 50 years and are now presumed to be extinct.

The introductory sections provide information about the physical environment, floral regions and biomes of South Africa. They also provide an overview of the orchid flora, the principles according to which orchids are named, and the structure, reproductive biology and conservation of orchids. We have purposefully omitted a section on cultivation, partly because this is a highly technical topic that is already covered in specialist literature, but mostly because this guide is intended specifically to enhance the appreciation of orchids in their natural habitat.

A population of the Cape orchid *Satyrium carneum* in coastal fynbos vegetation.

ACKNOWLEDGEMENTS

Herbert and Helga Stärker travelled hundreds of thousands of kilometres in search of orchid species to photograph for this field guide, relying on the generous assistance of many botanists, orchid enthusiasts and farmers along the way. Special mention must be made of the extraordinary support provided by Pat Brown, Shane Burns, John Burrows, Gareth Chittenden, Godfrey and Helene Coetzee, Hildegard Crous, Pieter Drake, Erica du Toit, Horst Filter, Graham Grieve, Lourens Grobler, Marcel Jansen, Kevin Joliffe, Judd Kirkel, Frans Krige, Bill Liltved, Mervyn Lotter, Gavin McDonald, Cameron McMaster, Douglas McMurtry, Warren Meyers, Geoff Nichols, Mike O'Connor, Ted Oliver, Tessa Oliver, Martin Rautenbach, John Rourke, Christo Smith, Roger Smith, Sarel Spies, Guy Upfold, Timo van der Niet, Jan Vlok, Martin von Fintel, Courtney Walton and Natasja Wortel.

Herbert Stärker also acknowledges the following for providing the additional photographs needed to illustrate aspects of some of the species in this guide: Uli Ade, Jill Blignaut, Benny Bytebier, Gareth Chittenden, Ruth Cozien, Hildegard Crous, Ezemvelo KZN Wildlife, Lourens Grobler, Karin Hansen, Steve Johnson, Judd Kirkel, Patrick Lane, Bill Liltved, Peter Linder, Mervin Lotter, Cameron McMaster, Douglas McMurtry, Geoff Nichols, Greg Nicolson, Marc Nicolson, Ted Oliver, Martin Rautenbach, Peter Swart, Ross Turner, Timo van der Niet, Jan Vlok, Louis Vogelpoel, Martin von Fintel, Karsten Wodrich, Peter Wragg, Bart Wursten and Helmuth Zelesny.

Steve Johnson wishes to thank Bill Liltved, with whom he worked for almost two decades on *The Cape Orchids* project, Craig Peter, Bruce Anderson, Timo van der Niet and the many other former and current members of his lab who have shared new discoveries of orchid biology; and Kathy Johnson for her vital role as an 'in-house' editor.

Benny Bytebier would like to thank Dirk Bellstedt, Bill Liltved, Peter Linder, Ted Oliver, Tessa Oliver and many other plant enthusiasts for introducing him to South African orchids; Kate and Graham Grieve for proofreading; and Jane Sakwa Makokha for her support and encouragement.

Peter Linder and Hubert Kurzweil granted us permission to use and modify the technical descriptions of South African orchid species that were included in their 1999 publication, *Orchids of Southern Africa*. We also acknowledge the help of Rafaël Govaerts and Denis Filer, the developers of the World Checklist of Selected Plant Families and the Botanical Research and Herbarium Management System (BRAHMS), respectively, as these two resources have greatly simplified the task of collecting and integrating information on South African orchids.

Pippa Parker at Struik Nature enthusiastically adopted this project. We have enjoyed working with her team of Helen de Villiers (Managing Editor), Sarah Gruft (Project Manager), Emsie du Plessis (Editor), Janice Evans (Design Director) and Tessa Fortuin and Neil Bester (typesetting and layout). Publication was made possible through generous sponsorships by the Botanical Society of South Africa, Elizabeth Parker of Elandsberg Nature Reserve and the Botanical Education Trust.

Polystachya pubescens growing in leaf litter on a sandstone boulder in southern KwaZulu-Natal.

INTRODUCTION

SOUTH AFRICAN ORCHIDS: AN OVERVIEW

Most people think of orchids as exotic blooms that hang from trees in tropical rainforests, or occasionally appear for sale in the local supermarket. To the uninitiated, it may therefore come as a surprise to learn that South Africa has close to 500 native orchid species. This is, of course, only a small fraction of the estimated 26,000 orchid species worldwide, but the orchid flora of South Africa is actually rich for a temperate country, exceeding that of the United States and Canada (213 species) and the whole of Europe (182 species). It is also notable that the majority of the orchids found in South Africa, Lesotho and Swaziland are endemic (restricted) to this region. Of the total of 473 orchid species in these three countries, 313 (66%) are found nowhere else.

As is the case in other temperate regions, about 90% of the orchids in southern Africa are terrestrial (growing in the ground) rather than epiphytic (growing on trees or shrubs) or lithophytic (growing on rocks). Owing to their underground tubers, terrestrial orchids are far better equipped than epiphytes or lithophytes to cope with the extremes of seasonal drought, low humidity, fire and winter frost that characterize most temperate regions. And, of course, epiphytes simply cannot exist in treeless habitats, such as the grassland biome that covers vast areas of South Africa.

The southern African region is world famous for its plant diversity. Almost 20,000 plant species are found in South Africa alone, and the greater Cape Floral Region with its estimated 9,000 plant species is considered so extraordinary that it has been recognized as one of only six plant kingdoms worldwide. Plants have undergone rampant diversification in southern Africa due to the wide variety of habitats in the region, ranging from humid coastal forests to craggy mountains. To survive and reproduce, plants in each of these diverse habitats must adapt to their unique soils, climate and pollinator fauna.

From the late seventeenth century onwards, orchid specimens were shipped from South Africa to Europe by various explorer naturalists. These collectors, notably Carl Thunberg, Anders Sparrman and Francis Masson in the eighteenth century and Johann Drège, Christian Ecklon and William Burchell in the nineteenth century, made numerous epic journeys across the southern African continent in search of new plant species. The process of formally describing and naming these specimens was carried out mainly by eminent European botanists of the eighteenth and nineteenth centuries, such as Carl Linnaeus and his students in Sweden, John Lindley in Britain, and Otto Sonder, Rudolf Schlechter and Heinrich Reichenbach in Germany.

The first guides to South African orchids intended for use by the general public (as opposed to specialist botanists) were produced in the late nineteenth and early twentieth centuries by Harry Bolus, a Cape Town stockbroker and amateur botanist. Bolus himself described more new South African orchid species than any other botanist, apart from the prodigious English botanist John Lindley. These extraordinary books, illustrated largely by Bolus's own paintings and drawings of floral dissections, are now valuable Africana and almost unobtainable. Bolus depicted about 300 South African orchid species in his books. The field guide *Wild Orchids of Southern Africa* (1982) by Joyce Stewart, Peter Linder, Ted Schelpe and Tony Hall covered not only South Africa but also the

Terrestrial orchids, such as *Pterygodium inversum*, thrive in seasonal environments.

Lithophytic orchids, such as *Disa comosa*, are specialized for growing on rocks.

A decaying tree festooned with a colony of the epiphytic orchid *Bulbophyllum sandersonii*.

neighbouring countries of Lesotho, Swaziland and Namibia, and listed 433 species. The reference work *Orchids of Southern Africa* (1999) by Peter Linder and Hubert Kurzweil covered the same territory, as well as Botswana, and listed 466 species. It is a testimony to the constant rate of discovery of new species that the present field guide, which covers only South Africa, Lesotho and Swaziland, lists 473 species. New orchid species continue to be discovered and described in South Africa at an average of about one species per year. We have no doubt that the tally of orchid species recorded for South Africa alone will eventually surpass 500, although, sadly, the actual number of species in the wild will inevitably decline as a result of human-induced extinction.

THE PHYSICAL ENVIRONMENT

South Africa is a country of extraordinarily diverse environments, ranging from high mountains through semi-arid plains to humid coastal regions. The most important physical feature is the Great Escarpment, which arcs around the country, separating a high-lying inland plateau from the lower-lying coastal zone.

In some places, notably the Cape Fold Mountains and the Drakensberg of KwaZulu-Natal, the escarpment forms a wall of peaks towering up to 3,500m, whereas in other places the drop-off is more gentle. The inland plateau, which includes Lesotho, is highest along the edge of the escarpment and, as a result, rivers drain into the Atlantic Ocean to the west of the watershed and into the Indian Ocean to the east.

Much of South Africa has an ancient, weathered surface with soils that are relatively poor in nutrients. The quartzitic soils of the Cape Fold Mountains, in particular, are extremely low in mineral nutrients, resulting in very slow plant growth. In the east of the country, short 'sourveld' grassland occurs on poor soils, often in high-rainfall regions. Many terrestrial plants, including orchids, rely on symbiotic associations between their roots and fungi to supplement their nutrition.

The other important feature that determines the distribution of vegetation in the region is the greatly contrasting patterns of rainfall. There is a gradient from winter rainfall in the west, based on the passage of frontal systems, to summer rainfall in the east,

The Drakensberg range, source of the Tugela River, is the highest and most dramatic section of the Great Escarpment.

Quartzitic soils are very poor in nutrients, but support a variety of orchids, such as *Disa cornuta*.

The Cape Fold Mountains had their origin in ancient geological upheavals, still evident in the leaning peaks of the Kammanassie range.

The rare orchid *Disa scullyi* is specialized for growing in damp montane grasslands.

based largely on thunderstorm activity. In addition, the total amount of rainfall is highest along the southern and eastern coastal belt, and lowest in the interior and along the western coastal belt.

FLORAL REGIONS

Botanists define floral regions (sometimes called botanical provinces) according to their concentrations of particular plant families and the extent to which plant families, genera and species are endemic (restricted) to those regions. The most notable floral region in South Africa is the Greater Cape Floral Region, which extends in a great L-shape from northern Namaqualand, through the south-western Cape, and as far east as Grahamstown. The Greater Cape Floral Region consists of approximately 9,000 plant species (almost half the total for South Africa), about 69% of which are endemic to the region. For these reasons, most botanists have recognized the Cape flora as one of the world's six floral kingdoms. There are 241 orchid species in the Cape Floral Region (again, about half of

the total for the whole of South Africa) and 68% of these are endemic. The Tongaland-Pondoland Floral Region on the east coast is mostly confined to South Africa and, like the Cape Floral Region, is considered a global botanical 'hot spot' on account of its combination of high species diversity, endemism and vulnerability to human-induced habitat transformation. The Eastern Mountain Floral Region, another botanical 'hot spot' with high levels of endemism, has strong affinities to both the Afromontane region of East Africa and the Greater Cape Floral Region. Three other floral regions – Karoo-Namib, Kalahari-Highveld and Zambezian – account for the remaining portion of the flora and also cover large areas of territory north of South Africa's borders.

BIOMES AND VEGETATION TYPES

Unlike floral regions, which are defined by plant species composition, biomes are defined by the structure of their vegetation. Biomes are determined largely by aspects of the physical landscape, most

Vegetation Biomes and Floral Regions of South Africa, Lesotho and Swaziland

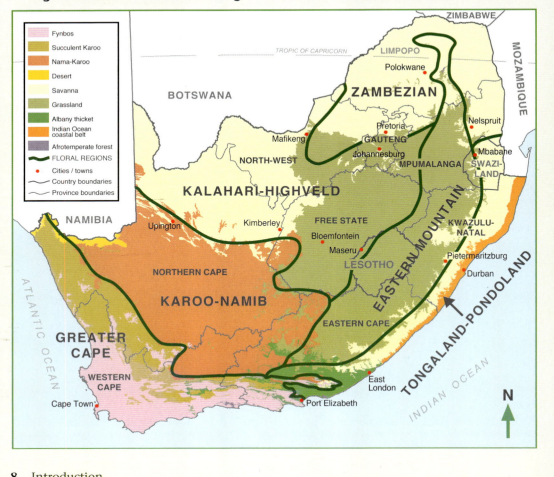

Fire consumes fynbos vegetation every 10–30 years, but stimulates many orchid species to flower.

Disa tripetaloides is one of several Cape orchids that occur only on stream banks.

The dainty orchid *Disa maculata* is found in rocky places that are draped in low-lying clouds in summer.

Disa rosea is specialized for growing on mossy cliffs in fynbos.

notably rainfall and soils. Each of the six biomes in South Africa thus represents a set of different ecological challenges that must be overcome by orchids. An understanding of the ecology of biomes is essential for the orchid enthusiast, as the distribution of the various orchid species largely corresponds to those of the biomes to which they have become specialized.

Fynbos

The vegetation of the fynbos biome is evergreen shrubland dominated by proteoid shrubs and ericoid heaths. It is the richest South African biome in terms of plant species – most of the roughly 9,000 plant species in the Cape Floristic Region are found in this vegetation type. Fynbos occurs mostly in the winter-rainfall region, although it also extends to the southern Cape where the rainfall is non-seasonal. Fynbos is typically found in the Cape Fold Mountains on quartzitic soils that are extremely poor in nutrients. Grasses are almost absent, their ecological niche being filled by the reedlike restios; where restios dominate the fynbos vegetation, this is known locally as 'restioveld'. Fires occur at intervals of 10–30

years. Whereas some fynbos plants survive fire by resprouting, many are in fact killed by fire but survive in the form of seeds to produce the next generation.

Fynbos biome orchid genera with the greatest number of species are *Disa*, *Satyrium*, *Holothrix*, *Pterygodium*, *Corycium* and *Acrolophia*. The genera *Ceratandra*, *Evotella* and *Pachites* are confined to the fynbos biome. The Cape Fold Mountains have a number of very distinctive habitats to which orchids have become specialized. These (with some examples of orchids that occur there) include mossy cliffs (*Disa longicornu* and *Disa rosea*), peaty marshes (*Corycium carnosum*, *Disa atricapilla*, *Disa racemosa*) and streamsides (*Disa tripetaloides*, *Disa uniflora*). Although fynbos has little agricultural value, it has been transformed in many places by invasive alien plants and urban development. As a result, many species, including dozens of orchids, are threatened with extinction.

Another type of shrubland vegetation found in the lowlands and in valleys in the fynbos biome is renosterveld. Unlike true fynbos, renosterveld occurs on relatively rich clay soils. Renosterveld supports a characteristic orchid flora, including orchids such as

Flat succulent leaves, such as those of *Holothrix secunda*, are a distinctive feature of many orchids found in the drier parts of southern Africa.

Satyrium pulchrum is restricted to a few granite outcrops in the succulent Karoo.

Disa spathulata, *Pterygodium catholicum* and *Satyrium erectum*. Since these soils are suitable for the production of wheat and grapes, much of the renosterveld has been claimed for agriculture. As a result, many of the orchids unique to this vegetation are now critically endangered.

Succulent Karoo

This biome is found mainly along the west coast, but narrow strips of succulent Karoo vegetation are also found in the valleys between the Cape Fold Mountains due to the rain shadow. The biome is characterized by low, shrubby vegetation dominated by vygies (Aizoaceae). Fires occur rarely and are not a key aspect of the ecology of plants in this biome. The west coast, particularly Namaqualand, is the richest semi-arid region worldwide in terms of plant diversity and is famed for its mass flowering of annual daisies in early spring. Rainfall is generally very low and occurs in winter. Given the low rainfall, it is surprising that several orchids, such as *Disa karooica*, *Disperis purpurata*, *Holothrix aspera* and *Satyrium pulchrum* occur in this biome; they generally have succulent leaves that lie flat on the ground.

Nama-Karoo

Vast areas of the interior are covered in a sparse layer of shrubs, most belonging to the daisy family (Asteraceae) – this is known as the Nama-Karoo biome. Rainfall is generally very low and occurs as sporadic thundershowers. The region is mainly utilized for sheep farming and is very poor in orchid diversity.

Grassland

Much of the higher-lying plateau and great escarpment of the central and eastern interior belongs to the grass-land biome, which is characterized by the absence of a shrub or tree layer. The grassland biome extends to some areas along the eastern coastal belt. Grasslands in South Africa, particularly those in the mountains and along the coast, are very rich botanically and contain the highest diversity of orchids outside of the

Namaqualand, part of the succulent Karoo biome, is famous for its displays of wildflowers in spring.

Cape Floral Region. The genera *Disa* and *Satyrium* are particularly well represented in this biome, along with *Habenaria*, *Eulophia*, *Disperis* and *Schizochilus*.

A very distinctive orchid habitat in this biome is wetland, which supports a plethora of specialized orchids such as *Disa rhodantha* and *Satyrium hallackii*. Higher-lying grasslands are prone to frost, which limits plant retention of leaves throughout winter. Grassland plants, including orchids, generally recover from fire (a frequent occurrence in this biome) and harsh winter conditions by sprouting from underground storage organs.

Grasslands have been highly transformed with those on richer soils being used extensively for agriculture, especially maize production and cattle grazing, while a significant portion of the higher-rainfall grasslands have been converted into exotic timber plantations. Grasslands in conservation and cattle-grazing areas are generally burnt every two years, or even annually, as a management practice.

Savanna

Savanna is the most common vegetation type in Africa and is well represented in the northern and eastern parts of South Africa. It is characterized by a grassy layer with scattered trees, notably acacias, which dominate the 'thornveld' type of savanna. Savanna is generally found in frost-free areas on rich soils and annual rainfall is typically low. Savanna plants are adapted to regular fires, which they survive either through fire-resistant bark or by resprouting from underground storage organs.

In terms of botanical diversity, savanna is not as rich as grasslands. Many of the orchids found here are not endemic to South Africa, but also occur further north. The dominant orchid genus in this biome in terms of number of species is *Eulophia,* followed by *Habenaria* and *Bonatea*. Characteristic epiphytic orchids in this biome are *Ansellia africana* and *Mystacidium capense*. Savanna is usually used for cattle and game farming and is relatively well conserved.

Afrotemperate forest

Evergreen afrotemperate forest is the smallest biome in South Africa. It occurs in scattered patches along the escarpment all the way from the Cape Peninsula to the Northern Province, usually in areas that are protected from fire. Interestingly, epiphytic orchids are not found in Cape forests west of Swellendam, presumably because humidity is too low during

the summers in this region. There are also many orchids specialized for life in the humus-rich forest floor. These include *Brownleea coerulea, Disperis fanniniae, Liparis bowkeri* and *Stenoglottis fimbriata*. Afrotemperate forests were heavily exploited for timber in the past, but are now relatively well protected.

Albany thicket

Thicket vegetation is found mostly in drier river valleys in the Eastern Cape, and generally consists of a tangled impenetrable mass of shrubs and succulents, especially *Aloe* and *Euphorbia*. Fire is not considered an essential part of the ecology of thicket. Thicket has a relatively low plant species diversity and is not particularly rich in orchids. A few *Eulophia* species and some epiphytes, such as *Mystacidium*, occur in this biome.

Indian Ocean coastal belt

This biome covers the largely subtropical climate zone of the east coast of the subcontinent. It includes the fabled Wild Coast, the iSimangaliso Wetland Park (formerly Greater St Lucia Wetland Park) and the

Montane grasslands in the Sehlabathebe National Park in Lesotho support a rich diversity of orchids.

Disa patula in grassland habitat in the foothills of the KwaZulu-Natal Drakensberg.

Large colonies of *Satyrium hallackii* subsp. *ocellatum* occur in wetland habitat in the grassland biome.

The grasslands of the Malolotje National Park in Swaziland contain many rare orchid species.

The most typical vegetation of the savanna biome consists of scattered thorn trees with an underlayer of grasses.

Afrotemperate forest in a sheltered ravine located in the KwaZulu-Natal Drakensberg.

Disperis fanniniae occurs in deeply shaded forest habitat along the eastern escarpment.

Kosi Bay lake system. The vegetation is mostly a mosaic of evergreen forest and fire-prone grassland. The high humidity and absence of a true dry season along the Indian Ocean coast belt mean that the forests support a rich flora of epiphytic orchids, many of which have ranges extending further north to the tropical areas of Africa. The grassland component of this biome also supports a rich orchid flora with distinctive species, such as *Disa caffra*, *Eulophia angolensis* and *Eulophia cucullata*. The biggest threats include coastal developments and mining of coastal sands for minerals.

ORCHID FLORA

With an estimated 26,000 species, orchids comprise the largest family of flowering plants. Most orchids are found in tropical America and Asia. There are, by our estimates, 471 orchid species that are native to South Africa. This number increases very slightly to 473 species when Lesotho and Swaziland are included. These are not big numbers in comparison to some countries in tropical America, such as Ecuador, which has more than 4,000 species. But to place the orchid flora of South Africa in perspective, one must compare it to those of other temperate regions, such as the United States and Canada (213 species) and Europe, extending to the Urals (182 species). Australia has 1,333 species, but this includes a significant number of tropical species found in the north of the country. As is the case in southern Africa, the orchid floras of other temperate regions are dominated by terrestrial orchids. What makes the orchid flora of South Africa truly remarkable is the high percentage of species that are endemic to the country.

There are five major orchid groups worldwide. These are essentially the main branches of the orchid evolutionary tree and are recognized by botanists as subfamilies. The oldest surviving branch (and the one with the most primitive characteristics) is the subfamily Apostasioideae, but this does not have any representatives in South Africa. The Vanilloideae, another group considered to have primitive characteristics, is represented in South Africa by a single species – *Vanilla roscheri*. The Cypripedioideae (slipper orchids) are also not represented in South Africa. The Epidendroideae are by far the richest subfamily worldwide in terms of species, and almost all of the epiphytic genera and some of the terrestrials (notably *Eulophia*, *Liparis* and *Orthochilus*) in South Africa belong to this group, which is characterized by its solid pollen masses (pollinia). The remaining subfamily is the Orchidoideae, a group to which most of the orchids in South Africa (as well as those in temperate America, Europe and Australia) belong. These orchids are almost always terrestrial and are usually characterized by pollen masses that are sectile (consisting of numerous subunits).

Botanical names

Botanical names of plants are known as binomials, as they are based on a unique combination of a genus and species name. A botanical name is only acceptable if it is accompanied by a formal description that could not apply to any other known plant species. *Disa uniflora*, for example, is the unique name for South Africa's most famous orchid. It was coined in 1767 by the Swedish botanist Peter Bergius, who was under the mistaken impression that plants of this species only ever produce one flower. It should be noted that the genus name is always capitalized, but not the species name. The name can also be abbreviated as *D. uniflora*, following the first mention of the full name. Occasionally, botanists will formally recognize geographical variation within a species by describing subspecies. For example, longer-spurred populations of *Satyrium stenopetalum* east of Swellendam have been described as subspecies *stenopetalum*, whereas shorter-spurred populations west of Swellendam have been described as subspecies *brevicalcaratum*. Varieties are similar, but do not have to be geographically separate. We also emphasize that the decision about whether or not to recognize different forms as distinct species is a matter of judgement. For example, some botanists recognize as many as eight *Stenoglottis* species, but here we recognize only four.

Few things frustrate the amateur plant enthusiast more than the propensity of botanists to change plant names. Although it may seem capricious, the renaming of plants is carried out according to a logical set of rules, with a bit of 'wiggle room' for interpretation. For example, botanists accept that a genus (and all other formal higher-level groupings, such as a tribe, family, order etc.) must contain all the species descended from a common ancestor. When it was recently discovered through molecular analysis that the *Eulophia* lineage (or branch of the evolutionary tree) contained not only all of the *Eulophia* species, but also all of the *Acrolophia* and *Oeceoclades* species, a decision was taken to place some *Eulophia* species into the genus *Orthochilus*, so that in the new classification, *Eulophia*, *Acrolophia*, *Oeceoclades* and *Orthochilus* each represent their own distinct lineages. In other cases, when faced with such situations, botanists opt to transfer all of the species from one genus to another. For example, all species in the genus *Schizodium* were transferred to *Disa* when it was realised that *Schizodium* species actually belong to the *Disa* lineage. Other cases are much harder to resolve. For example, it was recently suggested that all species of *Corycium* should be placed in *Pterygodium*, as species belonging to both genera

Albany thicket vegetation in the Eastern Cape.

Diminutive plants of *Holothrix parviflora* flowering beneath a giant *Euphorbia* in thicket vegetation.

The epiphyte *Mystacidium capense* is often abundant in thicket vegetation.

Large stands of *Eulophia angolensis* can still be found in moist grassland habitat along the Indian Ocean coastal belt.

Eulophia cucullata occurs in grassland habitat along the Indian Ocean coastal belt and further north into Africa.

Vegetation in the Indian Ocean coastal belt is typically a mosaic of grassland and forest.

were shown to be intermingled in the orchid evolutionary tree; however, there is uncertainty surrounding the genus name that should be applied to this merged group. This is because the name for a genus is derived from the identity of the 'type species' and, in this case, the type species for *Pterygodium – Pterygodium alatum* – does not seem to be closely related to the rest of the group. We therefore decided to keep the existing names for *Pterygodium* and *Corycium* species in this guide, even though they are likely to change again in the future.

Another reason for renaming species is the discovery that the specimen on which the original description was based actually belongs to another species that was named earlier. For example, in 1782 Carl Linnaeus's son described a species, *Disa grandiflora*, which was later realized to be the same as the *Disa uniflora* described by fellow Swede Peter Bergius in 1767. *Disa grandiflora* is therefore considered a synonym of *Disa uniflora*, as the older name generally has priority in botany, even though *Disa grandiflora* (the large-flowered Disa) seems more appropriate as a name. This situation is not uncommon, since botanists work in different countries or institutions and give new names to species that have already been described in another country or in a publication that they have not come across. For example, the wide-ranging species *Eulophia horsfallii* has no fewer than 25 synonyms! It can also arise when species are highly variable across their distribution, as is the case with *Satyrium bracteatum*, which has 11 synonyms. It should be noted that botanists are not under any obligation to accept the description of a 'new species'. If it is deemed that the 'new species' is not sufficiently distinct from an already-named species, then its name simply becomes a synonym of that established species. To make things easier for the reader, we have included synonyms in cases where

Forest encircling a pan in the Indian Ocean coastal belt is the habitat of the rare orchid *Vanilla roscheri*.

a binomial combination has changed since the publication of *Wild Orchids of South Africa* (1982).

We have not used common names in this book. The main reason is simply that, to the best of our knowledge, very few South African orchids have distinct common names. Instead, many names seem to refer broadly to groups of similar species. For example, the common name 'ewwa-trewwa' is given to a wide range of species in the genus *Satyrium*. The common name 'moederkappie' is used even more loosely to refer to a suite of *Disperis* and *Pterygodium* species with bonnet-shaped flowers.

Satyrium bracteatum is a highly variable orchid species; it has been given numerous names, all of which are now considered synonyms.

Disa uniflora plants frequently have more than one flower, despite the Latin name meaning 'one-flowered'.

Genus	Species in South Africa, Lesotho and Swaziland	Species endemic to South Africa, Lesotho and Swaziland	Species endemic to the Cape Floral Region
Disa	144	128	88
Satyrium	41	33	23
Habenaria	30	11	0
Eulophia	28	9	0
Disperis	26	18	6
Holothrix	23	19	7
Pterygodium	19	19	15
Corycium	14	12	8
Orthochilus	12	8	2
Polystachya	11	2	0
Bonatea	10	3	0
Schizochilus	8	7	0
Acrolophia	7	7	4
Mystacidium	7	6	0
Brachycorythis	7	1	1
Ceratandra	6	6	6
Brownleea	6	3	0
Angraecum	6	1	0
Huttonaea	5	5	0
Nervilia	5	0	0
Stenoglottis	4	2	0
Aerangis	4	0	0
Bulbophyllum	4	0	0
Oeceoclades	4	0	0
Tridactyle	4	0	0
Liparis	3	2	1
Microcoelia	3	0	0
Bartholina	2	2	0
Pachites	2	2	0
Cynorkis	2	1	0
Diaphananthe	2	1	0
Cyrtorchis	2	0	0

Orchid genera represented by two or more species in South Africa, Lesotho and Swaziland. Apart from these 32 genera, there are a further 22 genera represented by a single species.

Orchid genera represented in South Africa, Lesotho and Swaziland by a single species:

Acampe, Ansellia, Bolusiella, Calanthe, Centrostigma, Cheirostylis, Corymborkis, Didymoplexis, Dracomonticola, Evotella, Gastrodia, Jumellea, Margelliantha, Neobolusia, Oberonia, Platycoryne, Platylepis, Rangaeris, Rhipidoglossum, Vanilla, Ypsilopus, Zeuxine

Orchid genera that are endemic to South Africa, Lesotho and Swaziland:

Bartholina, Ceratandra, Dracomonticola, Evotella, Huttonaea, Pachites

Roots of *Stenoglottis longifolia* have a spongy velamen layer that absorbs water and nutrients.

Polystachya ottoniana is an epiphyte with a sympodial growth pattern – it produces a chain of pseudobulbs, each formed at the base of a shoot that eventually stops growing.

Cyrtorchis arcuata is an epiphyte with a monopodial growth pattern – the shoots keep growing indefinitely.

ORCHID STRUCTURE

Roots, stems and leaves (vegetative parts)

Orchids usually have simple, unbranched roots. In the epiphytic species, the roots creep over the bark of the host tree and absorb water and nutrients through a spongy layer known as the velamen. The velamen is usually white when dry and turns green when wet. Most orchids, particularly terrestrial ones, have symbiotic associations between their roots and fungi, which provide the plant with mineral nutrients and even carbohydrates from the very first stages of seed germination. Some orchids have no green leaves and rely entirely on fungi for supplying mineral nutrients and carbohydrates. Two examples featured in this guide are *Didymoplexis verrucosa* and *Gastrodia sesamoides*.

Thickened roots of orchids often serve as storage for carbohydrates and are known as tubers. This is particularly important for terrestrial orchids that are often dormant in a leafless state during dry periods of the year. Some of these root tubers can resemble testes and the name 'orchid' is in fact derived from the Greek word *orkhis*, meaning testicle. Orchid root tubers are replaced each year.

Many orchids store carbohydrates in swollen parts of the stem known as pseudobulbs. These are common among epiphytic orchids and are most pronounced in the genera *Bulbophyllum* and *Polystachya*, but are also found in some terrestrial orchids, notably *Eulophia* species. Most of the terrestrial orchid species in South Africa have simple, upright stems, which are replaced annually.

Leaves of these terrestrial species are usually soft in texture and are arranged spirally around the stem. Some *Holothrix* and *Satyrium* species that grow on the arid margins of the Karoo have thickened, succulent leaves that lie flat on the ground. Other orchids, notably *Acrolophia* species and most epiphytes, have leaves with a very hard, leathery surface that helps to prevent water loss. A few of the epiphytic orchids have leaves that are reduced to scales. Some of these, such as *Vanilla roscheri*, rely on their green stems for photosynthesis, whereas others, such as *Mystacidium gracile,* rely on their dense clusters of green, photosynthetically-active roots.

There are two types of growth forms and associated branching patterns in orchids. In most orchids, including all of the terrestrial species, growth is sympodial, meaning that a shoot stops growing when mature and is replaced by a new shoot which emerges from the base. Epiphytic orchids with this growth pattern, such as *Bulbophyllum* and *Polystachya*, often have several shoots adjacent to each other in a chain-like formation, frequently with pseudobulbs at the base of each stem. In a much smaller number of orchids, including *Cyrtorchis*, *Mystacidium* and *Tridactyle*, growth is monopodial, meaning that a shoot keeps growing indefinitely, and pseudobulbs are always absent.

Flowers and seeds

Orchid flowers are usually borne in groups known as inflorescences, although some species, such as *Bartholina burmanniana*, produce solitary flowers. The inflorescence is usually unbranched, but there are some exceptions, such as species of *Acrolophia* that have branched inflorescences. In epiphytes with sympodial growth, the inflorescence is either produced at the end of a stem (terminal position) or from its side (lateral position), whereas epiphytes with monopodial growth always produce inflorescences from the side of a stem. In most terrestrial species, the inflorescence is produced at the end of a single stem, but some species of *Disa* and *Satyrium* produce a 'fertile' (flowering) stem adjacent to a 'sterile' (non-flowering) stem.

Orchids have some of the most complex flowers of all plants and the details of their structure can be bewildering, even to experienced botanists. However, the basic structure of the orchid flower is essentially like that of any lily, in consisting of three outer sepals and three inner petals. One of the features that make orchid flowers unique is that one of their petals (the lip) is usually highly modified and serves as a landing platform and guide for pollinators. This specialization of the lip also makes orchid flowers bilaterally symmetrical, as opposed to the radial symmetry of most other lily-type flowers that have three identical petals. The lip starts as the uppermost petal during development, but in most cases as the flower matures, it twists around 180° so that the lip occupies the lowermost position of the flower. This process is called resupination and orchid flowers with the lip lowermost are referred to as 'resupinated'. The lip is usually modified to produce a spur, which can be as long as 20cm in the case of some *Bonatea* species. Pollinators typically place their tongue in the spur in search of nectar. The genus *Satyrium* is unique among all orchids in having two spurs derived from the lip, which is also unusual in being uppermost (non-resupinate) on the flower and shaped into a hood. In *Disa* and *Brownleea* species, it is the median sepal, rather than the lip, that is spurred and shaped into a hood.

In some orchids, the petals, and more rarely the sepals, can be highly divided. In *Acrolophia* and *Eulophia*, the surface of the lip is often sculptured with a prominent central crest (irregular ridge) and numerous papillae. In *Disa* species that were previously placed in the genus *Herschelia*, such as *Disa lugens*, the lip is often highly divided and even beard-like. Similarly, in *Bartholina* species, the lip is highly divided and spider-like. In *Bonatea* and *Habenaria* species, the lateral petals and the lip are often deeply divided, with some lobes even fused to the sepals. In a striking departure from the general tendency in orchids for the lip to be the showiest floral part, *Brownleea* flowers have a lip that is reduced to a tiny vestigial filament, and *Disperis* flowers have a lip that is mostly hidden within the flower.

In orchids, male and female reproductive structures, including the anther and stigma, are fused into a structure known as a column; the structure and shape of the column can be a valuable clue for identification. For example, species in the genera *Bonatea* and *Habenaria* often have a column with two distinct stigma lobes that protrude from the flower. In *Pachites* and *Satyrium*, the column is shaped like a pillar. In epidendroid orchids (of the subfamily Epidendroideae) the column usually has a distinctive anther cap.

If one is ever in any doubt as to whether a flowering plant is an orchid or not, the best clue lies in the structure of the pollen. Orchid pollen is not loose and powdery; instead, it is packaged into masses known as pollinia. These packages have a sticky pad (the viscidium), which becomes glued to the body of the pollinator. If you swipe your finger over a flower or prod it with a toothpick and a pollen package emerges, then it is almost certain that you have found an orchid, as only two plant families – the orchids and the milkweeds – have pollen that comes in the form of pollinia. Orchid pollinia are either inserted whole into a stigma cavity, or break up into small pieces that adhere to the surface of the sticky pad-like stigma.

The large pseudobulbs of *Eulophia petersii* enable this orchid to survive in seasonally dry savanna habitats.

Each plant in this colony of spider orchids (*Bartholina burmanniana*) produces a single flower.

The flowering shoot of *Satyrium longicauda* is accompanied by a much smaller non-flowering shoot that emerges alongside it.

Orchid pollen is packaged into pollinia. Here a *Cyrtothyrea* beetle carries the pollinia of *Eulophia parviflora*.

Wind disperses these dustlike seeds being released from the capsules of *Eulophia speciosa*.

Once fertilized by pollen, orchid flowers produce fruit containing thousands of dust-like seeds. These seeds are dispersed in the wind or very occasionally, as in the case of *Disa uniflora*, by floating on water. Orchid seeds are unusual among flowering plants in that they lack storage tissue and must rely on fungal associations to provide them with the nutrition required to develop into a seedling.

ORCHID REPRODUCTIVE BIOLOGY

Flowering times

The best months to see orchids in flower will differ between the winter- and summer-rainfall regions, and also between the coast and the mountains. In the winter-rainfall region, peak flowering for orchids is during October, November and December in the mountains, and during September in the lowlands. In the summer-rainfall region, flowering of orchid species in the higher-lying grasslands peaks in December and January, whereas in the grasslands of the Indian Ocean coastal belt and the lower-lying savanna regions, flowering tends to be earlier with a peak in October. Flowering of epiphytic orchids tends to be more evenly distributed throughout the year.

Response to fire

Many terrestrial orchids have become adapted to fire, which is an integral part of the ecology of many of South Africa's biomes. Indeed, some terrestrial orchids in the fynbos biome have developed a specialized strategy of flowering only in the first year after a fire. Incredibly, this means that such orchids may wait up to 30 years before flowering. It is believed that by flowering immediately after fire, these orchids gain the advantage of releasing their seeds into an environment that is ideal for seedling

establishment, as it is relatively free of competition from other plants. This flowering strategy is particularly evident among orchids that occur in marshes, as the vegetation in this habitat grows particularly rapidly and chokes out smaller plants. Some orchids in the fynbos biome will flower for two or three years after a fire and then cease flowering until after the next fire, whereas others, such as *Disa ferruginea*, will flower even in very mature fynbos vegetation.

Pollination (sexual reproduction)

In his 1862 book *On the various contrivances by which British and Foreign orchids are fertilised by insects, and on the good effect of intercrossing*, Charles Darwin showed convincingly that the complexity

A blue-banded *Amegilla* bee prepares to sip nectar from the flowers of *Disa versicolor*.

Fires in the Cape fynbos often trigger mass flowering of orchids, such as *Pterygodium acutifolium*.

Not all Cape orchids require fire to stimulate flowering. *Disa ferruginea* is usually seen flowering in mature fynbos vegetation.

of orchid flowers reflects various modifications for pollination. Many orchids lack nectar in their flowers and dupe their pollinators by mimicking the flowers of other plants. For example, *Disa pulchra* is an excellent mimic of the flowers of *Watsonia lepida* with which it grows in the grassland region. In general, pollinators make fleeting visits to orchid flowers and are not easy to observe.

The first discoveries of pollinators for South African orchids were in the late 1800s, when Stellenbosch chemist Rudolf Marloth discovered that the mountain pride butterfly pollinates the red-flowered orchids *Disa ferruginea* and *Disa uniflora*. To date, pollinators have been documented for more than 100 South African orchid species. Where information on pollinators is available, we have included it in the ecology notes for each species in this guide.

Bees are the most important pollinators of orchids and probably pollinate around 50% of South African orchid species. Some bees that play an important role (with examples of orchids they pollinate) include the large carpenter bees (*Disa racemosa*, *Eulophia speciosa*), blue-banded bees (*Disa versicolor*, *Satyrium sphaerocarpum*) and leafcutter bees (*Disa tenuifolia*, *Eulophia streptopetala*). A particularly important group is that of the oil-collecting bees, which pollinate oil-secreting orchid species in the genera *Ceratandra*, *Corycium*, *Disperis*, *Huttonaea* and *Pterygodium*. Orchid flowers pollinated by oil-collecting bees tend to be yellowish or white and have a pungent, soapy odour.

Wasp pollination of orchids is much less common in South Africa than in Australia. A particularly interesting case is *Disa sankeyi*, which is pollinated by spider-hunting wasps.

Many orchids with small, pungent flowers are pollinated by flies. Examples include *Disa obtusa*, *Satyrium bicallosum* and *Schizochilus zeyheri*. The particularly unpleasant 'rotten meat' smell

Disa pulchra (right) does not produce nectar, but attracts pollinators through mimicry of the nectar-producing species *Watsonia lepida* (left).

Disa uniflora is pollinated exclusively by the large mountain pride butterfly, *Aeropetes tulbaghia*, which has a predilection for red flowers.

A rather bizarre pollination system is exhibited by *Disa sankeyi* – it emits scent that attracts spider-hunting wasps.

The flowers of *Disa obtusa* have a pungent odour and are pollinated by bibionid flies.

Pterygodium dracomontanum is one of many South African orchid species that are specialized for pollination by oil-collecting *Rediviva* bees.

of *Satyrium pumilum* attracts carrion flies. Small scarab beetles pollinate the yellow or cream flowers of *Ceratandra grandiflora, Orthochilus ensatus, Orthochilus welwitschii* and *Satyrium microrrhynchum*.

It has been estimated that 50% of all African orchid species are pollinated by moths, although the figure for South African orchids is probably closer to 30%. Pollination by hawkmoths, large energetic insects that hover while feeding, is very common among epiphytic orchids with long-spurred white flowers, such as *Cyrtorchis arcuata* and *Mystacidium capense*. Similarly, many terrestrial orchids with long-spurred white or pale-green flowers, such as *Bonatea speciosa, Disa cooperi, Habenaria epipactidea* and *Satyrium longicauda,* are pollinated by hawkmoths. These insects fly for a short period after dusk, which explains why the orchids they pollinate often release a heady, sweet scent at that time of day. Smaller moths that settle while feeding have also been reported to pollinate orchids, such as *Satyrium bicorne,* that have pale flowers with relatively short spurs.

Orchids that occur in habitats that are not conducive to insect activity may self-pollinate. For example, *Disa glandulosa, Disa rosea* and *Disa vaginata* are all self-pollinating species that occur in shady, mossy rock clefts in the Cape Fold Mountains. The most infamous self-pollinating South African orchid is *Disa bracteata,* which has invaded large areas of Australia, presumably because its ability to self-pollinate allows it to produce large numbers of seeds without relying on insect availability. Another weedy, self-pollinating species that rapidly colonizes disturbed ground is *Disa woodii,* found in the eastern part of South Africa.

Vegetative (asexual) reproduction

Vegetative reproduction, which occurs through the proliferation of tubers at the end of long stolonoid (runner-like) roots, is found in some South African terrestrial orchids. Examples include *Disa harveyana, Pterygodium catholicum* and *Satyrium odorum*. The latter species is particularly prone to vegetative reproduction in areas where its oil-collecting bee pollinators are scarce or absent.

Hybridization

Orchids are famous for the ease with which they can hybridize. Worldwide there are more than 100,000 officially-named horticultural hybrids (known as grexes) resulting from artificial crosses among more than 5,000 orchid species and their hybrids. Even in the case of *Disa,* a genus containing just a few species that can be cultivated, more than 350 artificial hybrids have been recorded. The tremendous floral variation that can be obtained from hybridization is integral to the selective breeding of new orchid cultivars.

In their natural habitats, orchids are usually prevented from hybridizing – they may occur in different locations, flower at different times, have different pollinators, or place pollen on different parts of the body of pollinators, and thus are unlikely to exchange pollen. Nevertheless, pollinators do sometimes transfer pollen between different orchid species, giving rise to natural hybrids. These can be confusing to anyone trying to make an identification, particularly when the hybrids are backcrossed (when offspring are produced by cross-fertilization between hybrids and one or both of their parent species) resulting in a 'hybrid swarm' of plants that cover the full continuum of variation between species. The best-documented cases of hybrid swarms involve various Cape *Satyrium* species, notably *S. coriifolium* and *S. carneum* at some coastal sites, and *S. erectum*, *S. coriifolium* and *S. bicorne* at inland sites. Extensive hybridization and backcrossing can also occur between *Disa atricapilla* and *Disa bivalvata*, which share the same marsh habitats in the Cape Fold Mountains. In most cases, natural hybridization is a rare event and, even if the resulting hybrid plants reach maturity, they may not produce seeds if their flowers are too different from those of their parents to attract pollinators or place pollen on their bodies. Confirmed cases of natural hybridization are mentioned in the ecology notes for species in this book, and some of the most commonly encountered natural hybrids are illustrated on page 523.

ORCHID CONSERVATION

While researching this field guide, we became acutely aware of the dire conservation status of many of the South African orchid species. We would often arrive at a locality where a population had been well-documented in the past (for example, in herbarium records), only to find that the orchid had completely disappeared because the habitat had been totally transformed by urban development, invasive plants, livestock or cultivation of crops. It is a sobering fact that thousands upon thousands of orchid populations countrywide have been decimated by human-induced transformation of the environment. Some species have not been seen for more than a century and are considered to be extinct. The survival of many other species is hanging by a thread.

Official figures (the IUCN Red List of Threatened Species) prepared by the South African National Biodiversity Institute (SANBI) indicate that 70 species are threatened with extinction and a total of 140 species – roughly one quarter of the orchid flora – are of conservation concern. These figures are only approximate as there is very little data available, particularly for orchids that occur in remote localities or that flower only after fire and are thus seldom seen. Accurate monitoring of orchid

Hawkmoths pollinate many South African orchid species, including *Satyrium longicauda*, which attaches its pollinia to the tongue of these insects.

Disa woodii has bright yellow flowers, but does not depend on pollinators as it is capable of autonomous self-fertilization.

A hybrid (deep pink flowers) growing behind its parent plants, *Satyrium corriifolium* (orange flowers) and *Satyrium erectum* (light pink flowers), in the Western Cape mountains.

The critically endangered *Satyrium rhodanthum* is one of several grassland orchid species threatened by cattle grazing and commercial forestry operations.

populations is a valuable activity, which can be carried out by amateur orchid enthusiasts. One successful example is the long-term monitoring of *Disa barbata* on the Cape west coast by a team of botanists and amateur orchid enthusiasts. Much good work in this respect has been carried out in South Africa by CREW (Custodians of Rare and Endangered Wildflowers), a partnership between SANBI, the Botanical Society of South Africa and the KZN Biodiversity Stewardship Programme.

The greatest threat to orchid populations is habitat transformation, particularly due to urban development, agriculture (crop cultivation and farming with livestock), commercial forestry and alien plant invasions. Threats to different biomes vary – many of the documented extinctions or near extinctions in the Cape region have arisen due to urban development along the coast, crop cultivation and alien plant invasions; however, in the grasslands, commercial forestry and livestock farming are the major causes of habitat loss. Grazing by livestock, unless severe, does not always harm orchid populations, but conversion of grassland to pasture, as is the case with most dairy farming, leads to rapid loss of orchids especially if accompanied by the use of fertilizers. Direct collection of orchids from the wild, either by orchid enthusiasts or individuals harvesting for purposes of traditional medicine, has not yet threatened orchid populations to the same degree as transformation of their habitat. One exception is KwaZulu-Natal where the harvesting of orchids for traditional medicine occurs on a much larger scale.

In theory, the formally protected areas in South Africa, Swaziland and Lesotho should provide protection for many orchid species. In reality, most

of these protected areas were established for the conservation of large mammals and are situated in the savanna biome, which is relatively poor in orchid species diversity. There is a critical need for more formal protection of orchid-rich vegetation types, such as renosterveld and lowland fynbos in the Cape region, and montane grasslands in the eastern part of South Africa. Existing nature reserves, such as Verloren Vallei in Mpumulanga, Malolotje in Swaziland and Sehlabathebe in Lesotho, play a vital role in the protection of grassland orchid species.

Orchids worldwide have special status in terms of the Convention on International Trade in Endangered Species of Wild Fauna and Flora (CITES). According to this international agreement, orchids listed in Appendix II (which includes all of the South African species) may not be shipped between countries without a special permit. Furthermore, orchids are protected by the laws and ordinances of the various provinces in South Africa, according to which they may not be collected in the wild without a permit, and such permits are generally issued only for scientific studies.

Apart from its being illegal, there is very little point in digging out orchids and trying to cultivate them, as it generally leads to the death of the plant and robs other people of the opportunity to enjoy the plant in its natural habitat. Individuals interested in growing orchids should contact their local orchid society, and can purchase exotic hybrids or even take on the challenge of growing South African orchid species from seeds obtained from nursery-grown plants. However, nothing, in our opinion, surpasses the thrill of discovering and photographing wild orchids in their natural habitats.

FLORAL PARTS OF ORCHIDS

The labelled photographs on these pages illustrate the typical structure of flowers of various genera.

Orchidoid orchids

Bonatea

- upper petal lobe
- median sepal
- lower petal lobe
- anther
- rostellum
- stigma
- lip lobes
- lateral sepal
- ovary
- spur

Habenaria

Satyrium

- lip
- stigma
- anther
- petal
- lateral sepal
- median sepal
- spurs

Disa

- spur
- median sepal
- petal
- anther
- rostellum
- stigma
- lateral sepal
- lip

Disperis

- lip claw
- spur

Huttonaea

- median sepal
- petal
- lip claw
- lip appendage
- lateral sepal
- rostellum
- lip blade

Corycium

Epidendroid orchids

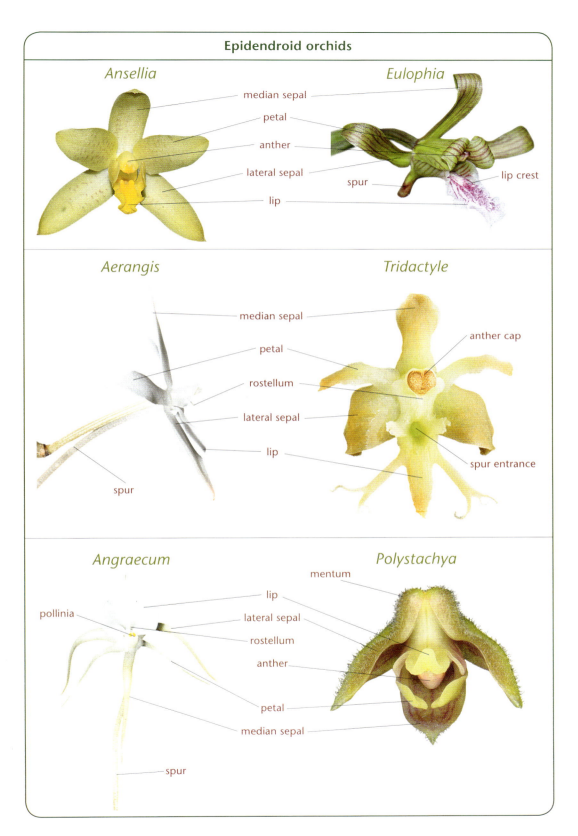

Ansellia
- median sepal
- petal
- anther
- lateral sepal
- lip

Eulophia
- spur
- lip crest

Aerangis
- median sepal
- petal
- rostellum
- lateral sepal
- lip
- spur

Tridactyle
- anther cap
- spur entrance

Angraecum
- pollinia
- lip
- lateral sepal
- rostellum
- anther
- petal
- median sepal
- spur

Polystachya
- mentum

Species in the orchid flora of South Africa that are not featured in this field guide, either because they have not been seen for many decades, or because no suitable illustrative material could be located.

Species	Last confirmed sighting
Corycium tricuspidatum	1977
Disa brevipetala	1942
Disa cedarbergensis	1987
Disa ecalcarata	1947
Disa forcipata	1897
Disa forficaria	1966
Disa hircicornis	1936
Disa newdigateae	c.1931–1935
Disa welwitschii	c.1931
Habenaria stenorhynchos	1894
Habenaria ciliosa	2014
Holothrix culveri	1890
Holothrix macowaniana	1892
Holothrix micrantha	1925
Nervilia renschiana	Unconfirmed
Oberonia disticha	2010
Pterygodium connivens	1976
Pterygodium newdigateae	c.1895

Left: *Disa forficaria* was last seen in 1966 in the mountains of the Western Cape.
Right: *Disa newdigateae* is considered to be extinct as the last sighting of this orchid was during the late nineteenth century.

HOW TO USE THIS BOOK

This guide is intended to facilitate the identification of orchids in their natural habitat and provide the user with some information regarding their ecology.

We use the currently accepted scientific name for each orchid species. The genera are listed alphabetically and species are arranged alphabetically within each genus. If the species was featured under a different name in *Wild Orchids of Southern Africa* (1982), *Orchids of Southern Africa* (1999), *Field Guide to the Orchids of Northern South Africa and Swaziland* by Douglas McMurtry, Lourens Grobler, Jolisa Grobler and Shane Burns (2008), or *The Cape Orchids: A Regional Monograph of the Orchids of the Cape Floristic Region* by Bill Liltved and Steve Johnson (2012), then this alternative name (the synonym) is given after the current name. Both the current name and the synonym are listed in the index.

Orchid identification is much easier when the search for a name can be narrowed down to the genus level. Genera tend to have quite distinctive features. For example, *Satyrium* is the only orchid genus with two floral spurs. We therefore encourage readers to familiarize themselves with the section on the basic structure of orchid flowers (in conjunction with the accompanying labelled photographs of orchid flowers) and the brief generic descriptions (in conjunction with photographs of representative species) that are provided in the introductory section to this guide. As an additional tool for identification, we provide a technical key to the genera at the back of the guide.

We have designed this guide so that readers should be able to identify particular species using the photographs, descriptions, maps and flowering-time bars in the species accounts. When selecting photographs, we have tried to illustrate the whole plant in its habitat, the inflorescence, and close-ups of the flowers from both front and side views, with particular attention to diagnostic characters. Different colour forms of flowers and close-ups of leaves are also given in cases where these could be useful for identification. The place and date of each photograph is given below the images. The maps show the extent of the known distribution of each taxon, shaded pink if they are endemic to the region covered by the guide and green if they are more widespread, but it should be understood that not all of the shaded area is necessarily suitable habitat for the taxon; in addition, most orchids have patchy distributions, with some populations separated by hundreds of kilometres. Similarly, the calendar bar shows the range of recorded flowering times, but individual plants may occasionally be found in flower outside of these typical periods.

A brief description is provided to confirm tentative identifications based on the visual aids. Parts of the plant are described from the stem up to the inflorescence, and from the outer parts of the flower to the inner parts. Descriptions of each flower part, such as the lip, are separated by semi-colons, and cover the shape and then size of the part. In cases where similar species could be confused, we have highlighted their most obvious differences in a section of text following the description. Where more than one subspecies or variety are found in the region covered by this guide, they are listed separately after the description of the species, together with an explanation of how to distinguish between them. In cases where only one subspecies or variety of a species is found in the region, this is indicated in the text below the description of the species. We have tried to keep specialized botanical terminology to a minimum, but where use of a term is unavoidable, an explanation of its meaning is provided in the glossary at the end of the book.

The ecological notes give an indication of the abundance of the species, its particular habitat, altitudinal range, pollinators (if known), natural hybrids, and whether fire is required for flowering.

Species name — Place and date of photograph — Synonym

Ecological notes — Distribution map — Calendar bar showing flowering season

ORCHID GENERA OF THE REGION

Acampe

A genus of eight species, mostly centred in Asia, with a single species in Africa.

Robust, monopodial epiphytes with a stout stem. Leaves in 2 opposite rows, leathery to slightly succulent, folded together lengthwise. Inflorescences compact, usually branched, arising from among the leaves, 1–many-flowered. Flowers non-resupinate, small to medium-sized, fleshy. Sepals and petals similar, free. Lip with a pouch, side lobes usually small.

Acrolophia

A genus of seven species, endemic to South Africa and largely centred in the Cape Floral Region, with one species extending into KwaZulu-Natal.

Terrestrial herbs with short rhizomes and fleshy roots; stems without pseudobulbs. Leaves in 2 rows, often forming a fan, folded lengthwise, leathery. Inflorescence

terminal, simple or branched, laxly to rarely densely few- to usually many-flowered. Flowers resupinate or not, medium-sized. Sepals and petals similar, free. Lip more or less 3-lobed, with dense to sparse papillae; spur short, usually present.

Acrolophia is closely related to *Eulophia*, but distinguishable by the terminal inflorescence, the 2-ranked leaves and the lack of pseudobulbs and corms.

Aerangis

A genus of 57 species largely endemic to Africa, with the exception of one species, which extends to Sri Lanka. Four species occur in South Africa.

Epiphytic, monopodial herbs with a short, unbranched, woody stem. Leaves in 2 rows, strap-shaped and widening towards the tip, unequally

bilobed, fleshy to leathery, usually with noticeable net-like venation. Inflorescences 1 to several, arising from among or below the leaves, few- to many-flowered. Flowers white, star-shaped. Sepals and petals similar, free. Lip usually wider than sepals and petals, with an elongated nectar-producing spur.

Angraecum

A genus of 222 species largely endemic to Africa, with the exception of one species, which is found in Sri Lanka. Its centre of diversity is in Madagascar where 138 species occur. There are six species in South Africa.

Epiphytic, monopodial herbs with short or long stems. Leaves in 2 rows, unequally bilobed at the tip, very variable in size. Inflorescences arising from among or below the leaves, 1–many-flowered. Flowers very variable in size. Sepals and petals similar, free. Lip distinctly concave and spurred. Rostellum divided into 2 broad lobes.

Ansellia

A genus with only one species, which is endemic to Africa.

Epiphytic herbs with upward-pointing, basket-like roots and erect, cane-like stems up to 1m tall. Leaves alternate, folded, thin but tough, on upper half of stem, lost with

age. Inflorescence terminal, branched, laxly many-flowered. Flowers large, showy and very variable in size, shape and coloration throughout its range, green to yellow and usually prominently spotted with brown to maroon. Sepals and petals similar. Lip 3-lobed, with 2 or 3 callus ridges on the midlobe.

Bartholina

A genus of two species confined to the Western, Eastern and Northern Cape, and southern Namibia.

Terrestrial herbs. Leaves solitary, round, lying flat on the ground. Inflorescence single-flowered; flower stalk minutely hairy, leafless. Flowers with petals and lip white to mauve, sepals

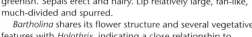

greenish. Sepals erect and hairy. Lip relatively large, fan-like, much-divided and spurred.

Bartholina shares its flower structure and several vegetative features with *Holothrix*, indicating a close relationship to that genus.

Bolusiella

A genus of four species endemic to Africa, with one species occurring in South Africa.

Very small, monopodial epiphytes; stemless or with very short, woody stem. Leaves succulent, usually

laterally compressed. Inflorescences arising from among or below the leaves, many-flowered; bracts largely obscuring the flowers. Flowers non-resupinate, small, white. Sepals and petals similar. Lip simple or 3-lobed; spur cylindrical.

Bonatea

A genus of 13 species largely endemic to Africa, with the exception of one species, which extends into the Arabian Peninsula. There are 10 species in South Africa.

Robust, terrestrial herbs. Leaves at the base or along the stem, sometimes withered at flowering time. Inflorescence terminal, unbranched, lax to dense, 2–many-flowered. Flowers resupinate, green and white. Sepals unequal; median sepal free, forming a hood with upper petal lobes; lateral sepals reflexed, united at the base with the lower petal lobes, the lip and the stigmatic arms. Petals deeply 2-lobed. Lip 3-lobed from a short, narrow base; spur with a tooth in front of the spur mouth.

Bonatea is similar to *Habenaria*, but is distinguishable by the hood-like central rostellum lobe and the presence of a tooth in front of the spur mouth.

Brachycorythis

A genus of 36 species, of which 24 are found in Africa and 12 in Asia. Seven species occur in South Africa.

Terrestrial herbs with an erect, often densely leafy stem. Leaves cauline, spear-shaped. Inflorescence terminal, with flowers subtended by bracts similar to the cauline leaves, and partially obscuring lowermost flowers, many-flowered. Flowers resupinate. Sepals free. Petals fused to the column at the base. Lip projecting forward, with the limb boat-shaped or spurred, and the blade flattened, entire or 2- or 3-lobed.

Brownleea

A genus of eight species endemic to Africa, with six species occurring in South Africa.

Slender, terrestrial herbs. Leaves 1–3 along the stem. Inflorescence a dense head or a lax spike, 1–many-flowered. Flowers white to mauve, often spotted with darker mauve. Median sepal fused with the petals to form a spurred hood; lateral sepals free. Lip minute, erect in front of the stigma.

Due to the spurred median sepal, *Brownleea* is superficially similar to *Disa*, but is distinguishable by the petals, which are fused to the median sepal, and the erect, almost obscure lip.

Bulbophyllum

A pantropical genus consisting of 1,877 species, of which 291 occur in Africa and four in South Africa.

Epiphytic, or sometimes lithophytic, sympodial

herbs on long or short, creeping and often branched rhizomes; pseudobulbs spheroidal to conical, often sharply angled, clustered or well-spaced, with 1 or 2 terminal leaves. Leaves leathery to fleshy. Inflorescence lateral from the base of the pseudobulb, 1–many-flowered, erect or hanging, usually with sessile flowers on opposite sides of the stalk or rarely with all flowers arising from a single point; flower-bearing part of the stalk often swollen and flattened. Flowers usually small, often fleshy. Sepals unequal; lateral sepals fused to the column foot to form a chin-like extension. Petals smaller than sepals, linear. Lip much smaller than sepals, hinged, curved, entire, without a spur. Pollinia 4, in pairs.

Calanthe

A genus of 209 species, centred mainly in Asia. Three species occur in Africa and one in South Africa.

Terrestrial herbs with a short, leafy stem,

obscurely swollen into a pseudobulb. Leaves usually large, pleated. Inflorescence simple, arising from among the leaves, many-flowered. Flowers often large and showy. Sepals and petals similar. Lip spurred, fused at the base to the column, with a distinct callus or crest near the base, 3-lobed with the midlobe usually 2-lobed.

Centrostigma

A genus of three species, restricted to continental Africa south of the Equator, with one species occurring in South Africa.

Robust, terrestrial herbs with a leafy stem. Leaves linear to spear-shaped, grading apically into the floral bracts. Inflorescence terminal, lax, 4–20-flowered. Flowers resupinate, medium-sized to large, white, cream or green. Sepals unequal; median sepal hooded; lateral sepals spreading. Petals similar to the lateral sepals, but smaller. Lip united with the column at the base, spurred, deeply 3-lobed with the side lobes fringed.

Centrostigma is very similar to *Habenaria*, but is distinguishable by the longitudinal division of the stigmatic arms.

Ceratandra

A genus of six species, endemic to the western and southern parts of the Cape Floral Region of South Africa.

Terrestrial herbs with slender to fairly robust stems. Leaves numerous, linear, forming a basal rosette as well as spaced along stem. Inflorescence terminal, lax to dense, 6–many-flowered. Flowers resupinate or not. Median sepal fused to the petals; lateral sepals concave. Lip spatula-shaped, with or without appendage; blade more or less anchor-shaped, semi-circular, often with a small callus.

Cheirostylis

A genus of 52 species centred mainly in Australasia and tropical Asia, with three species occurring in Africa and one in South Africa.

Terrestrial herbs with short, erect stems arising from a creeping base rooting at the nodes. Leaves thin-textured. Inflorescence terminal, dense, hairy, many-flowered. Flowers small. Sepals, petals and lip joined to form a funnel-shaped flower. Lip with a pouch at the base; apex more or less 2-lobed.

Corycium

A genus of 14 species restricted to South Africa with the exception of two species, which extend to Malawi and southern Tanzania.

Slender to robust terrestrial herbs. Leaves few to many, on the stem. Inflorescence usually dense, many-flowered. Flowers resupinate or very rarely not. Median sepal almost always fused to the petals and forming a hood; lateral sepals free to fused at the base. Lip fused basally to the column, bearing an appendage that usually overarches the column to a varying extent.

Corymborkis

A genus of six species, of which two occur in Africa, one in South Africa.

Terrestrial herbs with short, branching rhizomes and erect, reed-like stems up to 2m tall. Leaves arranged spirally, pleated, dark green. Inflorescences arising from among the leaves or terminal, few- to many-flowered. Flowers white to greenish-white, often fragrant. Sepals, petals and lip similar and very narrow, only the lip widening towards the tip.

Cynorkis

A genus with 156 species endemic to Africa. Its centre of diversity is in Madagascar, where 103 species occur. Two species are found in South Africa.

Terrestrial herbs, often with glandular-hairy stems. Leaves 1–few, basal. Inflorescence lax or dense, 1–many-flowered. Flowers resupinate, small. Sepals free, unequal. Petals erect, smaller than the sepals, often converging with the median sepal to form a hood. Lip entire, 3- or 5-lobed, spurred at the base.

Cyrtorchis

A genus of 18 species endemic to Africa, centred mainly in West and Central Africa. Two species occur in South Africa.

Epiphytic, or more rarely lithophytic, monopodial herbs with elongated, woody, usually erect, sometimes branched stems. Leaves strap-shaped, leathery to slightly succulent, equally or unequally bilobed, in 2 opposite rows. Inflorescences arising from among the leaves, arching, 5–15-flowered, with conspicuous bracts. Flowers resupinate, white, turning yellow or orange with age, sweet-scented. Sepals and petals similar, strongly recurved at the tips. Lip with a spur that is broader at the mouth and tapering towards the apex.

Diaphananthe ·

A genus of 33 species endemic to Africa with two species occurring in South Africa.

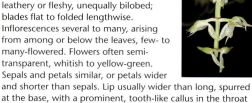

Epiphytic, monopodial herbs with woody, erect to hanging stems. Leaves leathery or fleshy, unequally bilobed; blades flat to folded lengthwise. Inflorescences several to many, arising from among or below the leaves, few- to many-flowered. Flowers often semi-transparent, whitish to yellow-green. Sepals and petals similar, or petals wider and shorter than sepals. Lip usually wider than long, spurred at the base, with a prominent, tooth-like callus in the throat of the spur. Column stout with both pollinia attached to a common viscidium by separate stipes.

Diaphananthe is similar to *Rhipidoglossum*, but is distinguishable by the conspicuous tooth in the mouth of the spur and by the pollinia, which are both attached to a single viscidium by separate stipes.

Didymoplexis

A genus of 17 species, centred mainly in Asia and the Pacific islands. Two species occur in Africa, one in South Africa.

Slender terrestrial herbs lacking chlorophyll, with erect, unbranched stems. Leaves reduced to scales. Inflorescence terminal, 1–many-flowered. Flowers non-resupinate, white to cream, short-lived, unspurred, with flower stalk elongating rapidly after pollination. Sepals and petals fused shortly at the base. Lip free, often lobed.

Disa

A genus of 183 species, with its centre of diversity in the Cape Floral Region. Largely endemic to Africa, with the exception of one species, which extends into the Arabian Peninsula. There are 143 species in South Africa and one species, *Disa intermedia*, is endemic to Swaziland.

Terrestrial, rarely epiphytic or lithophytic herbs, occasionally producing sterile shoots. Leaves along the stem or at the base, green at flowering time, rarely developing before or after flowering. Inflorescence simple, sometimes head-like, 1–many-flowered. Flowers usually resupinate. Median sepal hooded and spurred. Lip usually narrow.

A large and variable genus that has been expanded over the last few decades through the inclusion of the genera *Herschelia(nthe)*, *Monadenia* and *Schizodium*, which were shown to be part of the *Disa* lineage.

Disperis

A genus of 78 species almost completely confined to Africa, with the exception of one species, which occurs in Asia. There are 26 species in South Africa.

Small and slender terrestrial herbs. Leaves 1–5, sometimes reduced and sheathing, alternate or in a single opposite pair, arising from the base or along the stem, sometimes purple beneath. Inflorescence often 1-sided, 1–many-flowered. Flowers resupinate. Median sepal fused to the petals and often forming a hood or spur; lateral sepals spreading, each with a spur or pouch. Lip ascending inside the hood and usually consisting of a narrow claw, lip blade and an appendage of varying shape.

Some species are difficult to identify as the most distinctive character, namely the lip and its appendage, is hidden within the hood or spur.

Dracomonticola

A genus consisting of a single species restricted to the Drakensberg region of the Eastern Cape, KwaZulu-Natal and Lesotho.

Small and slender lithophytic or terrestrial herbs. Leaves solitary at the base. Inflorescence almost head-like, slightly nodding, few-flowered. Flowers with sepals similar; petals about half as long as sepals. Lip broadly spear-shaped, with obscure side lobes and a slight pouch at the base.

Eulophia

A genus of 167 species with its centre of diversity in Africa, where 125 species occur, 28 of which are found in South Africa.

Terrestrial herbs with leaf-bearing shoot next to flower-bearing stem. Leaves absent to fully developed at flowering time, thin-textured to sometimes leathery or succulent. Inflorescence lateral, erect, simple or rarely branched, few- to many-flowered. Flowers usually resupinate at flowering time. Sepals similar, free. Lip usually 3-lobed, with crests of lamellae, papillae or low, rather warty ridges, with or without a basal sac or a conical to cylindrical spur.

The closely related genus *Orthochilus* is comprised of species that were, until recently, included in *Eulophia*.

Evotella

A genus consisting of a single species endemic to the Western Cape.

Terrestrial herbs. Leaves linear to spear-shaped, cauline. Inflorescence fairly dense, many-flowered. Flowers resupinate, cup-shaped. Median sepal and petals fused; lateral sepals spreading. Lip with undivided, 2-lobed, elongate appendage.

Gastrodia

A genus of 65 species, centred mainly in Asia. Six species occur in Africa, with one species naturalized in South Africa.

Terrestrial herb lacking chlorophyll, with thick, fleshy rhizomes. Leaves reduced to small, membranous scales. Inflorescence lax, 1- to many-flowered. Flowers resupinate or not, on a distinct stalk, often nodding. Sepals and petals mostly fused into a tube, free lobes much shorter than the tube. Lip free, 3-lobed, mobile.

Habenaria

A genus of 839 species, distributed throughout the world's tropical and subtropical regions, with 239 species occurring in Africa and 30 occurring in South Africa.

Slender to robust terrestrial herbs. Leaves 1 or 2 and then flat on the ground or more often many spaced along the stem, with lowermost 2 or 3 smaller, those above much larger and then grading into floral bracts. Inflorescence terminal, lax to dense, few- to many-flowered. Flowers almost always resupinate, mostly green, yellowish-green or green and white. Sepals free. Petals entire or 2-lobed. Lip 3-lobed or less often undivided, with a long or short spur. Stigma split in two and often on conspicuously forward-projecting processes.

Holothrix

A genus of 46 species, largely endemic to continental Africa, with the exception of two species, which extend into the Arabian Peninsula. There are 23 species that occur in South Africa.

Slender terrestrial or lithophytic herbs. Leaves 1 or 2, round or egg-shaped, flat on the ground, sometimes withered before flowering time. Inflorescence erect, simple, usually hairy, with or without bracts, usually 1-sided or more or less 1-sided. Flowers small. Sepals similar, sometimes fused. Petals entire or fringed. Lip spurred, usually 3–many-lobed.

Holothrix shares its flower structure and several vegetative features with Bartholina, indicating a close relationship to that genus.

Huttonaea

A genus of five species endemic to South Africa and Lesotho.

Terrestrial herbs. Leaves 2, upper leaf often smaller. Inflorescence lax, 1–25-flowered. Flowers resupinate, white to pale green, often marked with purple. Median sepal erect; lateral sepals larger and spreading. Petals larger than sepals, with a

claw at the base, expanding above into a fringed blade. Lip spreading, broad and conspicuously fringed, without a spur.

Jumellea

A genus of 60 species endemic to Africa, with its centre of diversity in Madagascar. Only one species occurs in South Africa.

Epiphytic or lithophytic, monopodial herbs with upright, woody stems, branching and bearing roots only near the base. Leaves

leathery, unequally bilobed at the tip, arranged in 2 rows. Inflorescences arising from among the leaves, single-flowered. Flowers large, white. Sepals and petals similar, narrow, reflexed to spreading. Lip diamond-shaped; claw with a central ridge; spur slender.

Liparis

A genus of 425 species with a worldwide distribution. There are 69 species in Africa, of which 43 occur in Madagascar and the Western Indian Ocean islands, and three in South Africa.

Terrestrial or epiphytic herbs, usually with pseudobulbs. Leaves 1–4, more or less pleated and usually thin-textured. Inflorescence erect, terminal, few- to many-flowered. Flowers resupinate, mostly rather small, green to lime, becoming orange with age. Median sepal much narrower than lateral sepals, which are united in the lower half behind the lip. Petals reflexed, often linear. Lip simple or 2-lobed, firmer than sepals.

Margelliantha

A genus of six species endemic to continental Africa, with one species occurring in South Africa.

Epiphytic, monopodial herbs with short stems. Leaves leathery, in 2 opposite rows, unequally bilobed at tip. Inflorescences lateral, arising

from among the leaves, 3–12-flowered. Flowers bell-shaped, white to yellow or pale green. Sepals and petals free. Lip concave, lacking a callus, spurred at the base, with the spur spherical or pouch-shaped.

Microcoelia

A genus of 31 species endemic to Africa, with three species occurring in South Africa.

Leafless, epiphytic, monopodial herbs, usually with conspicuous, often long, branched roots. Stem usually very short but occasionally long. Inflorescences concentrated on apical part of stem, few- to many-flowered. Flowers small to minute, white, often with a green to orange tinge.

Sepals and petals similar, free to fused at the base. Lip free, entire to 3-lobed; spur spherical to cylindrical.

Mystacidium

A genus of 10 species endemic to Africa, with seven species occurring in South Africa.

Small, epiphytic, monopodial herbs with very short stems and many roots. Leaves in 2 opposite rows, strap-shaped, leathery. Inflorescences

several, simple, arising from among or below the leaves, 1–13-flowered. Flowers resupinate, small, white or pale green. Sepals similar. Petals smaller than sepals. Lip 3-lobed, without a callus; midlobe sharply deflexed, with a long, tapering, often curved spur.

Some species cannot easily be distinguished solely on morphological characters and can be separated by flowering times only.

Neobolusia

A genus of three species endemic to continental Africa, of which one species occurs in South Africa.

Slender, terrestrial herbs. Leaves 1–3 near the base and several sheathing leaves on the stem, grading into the floral bracts. Inflorescence simple, lax to fairly dense, 1–12-flowered. Flowers resupinate, few to many. Sepals and petals similar. Lip with a callus at the base and wavy edges, without a spur.

Nervilia

A genus of 69 species found in the tropics and subtropics of Africa, Asia and Australasia, with 17 species in Africa, five in South Africa.

Terrestrial herbs. Leaves solitary, heart-shaped or almost circular, stalked, pleated. Inflorescence simple, 1–many-flowered. Flowers resupinate. Sepals and petals similar, free. Lip 3-lobed in the basal part, entire or fringed, variously keeled or crested on upper surface, without a spur.

The various species can be difficult to identify as leaves and flowers do not generally occur at the same time.

Oeceoclades

A genus of 40 species with its centre of diversity in Madagascar. Largely endemic to Africa, with the exception of one species, Oeceoclades maculata, which has become a weed in tropical parts of the Americas. Three species are found in

South Africa and one species, Oeceoclades quadriloba, is found in Swaziland.

Terrestrial or rarely epiphytic, sympodial herbs with aerial pseudobulbs. Leaves 1–3, mostly folded lengthwise, usually with a stalk, often variegated. Inflorescence lateral, simple or branched, few- to many-flowered. Flowers resupinate, small to medium-sized. Sepals and petals similar, free. Lip 4-lobed, sometimes obscurely so, with crest of transverse or longitudinal ridges at the entrance to the spur.

Orthochilus

A genus of 34 species, of which 32 occur in Africa, 12 in South Africa.

Terrestrial herbs with leaf-bearing shoot next to flower-bearing stem. Leaves absent to fully developed at flowering time, pleated, thin-textured to sometimes leathery. Inflorescence lateral, simple, usually dense to almost spherical, usually many-flowered; bracts persistent, often conspicuous. Flowers mostly nodding, not opening widely, uniformly coloured, sometimes with conspicuous, differently coloured disc or papillae on the lip. Sepals and petals similar. Lip 3-lobed, with papillose ridges in basal half, often ending in scattered papillae or warts on the midlobe; spur cylindrical to club-shaped, or absent.

Similar to Eulophia, and previously included as part of that genus, but distinguishable by the usually head-like inflorescence with bell-shaped, nodding flowers, by the petals and sepals that are similar in size, shape and colour, and by the papillate distal crest of the lip.

Pachites

A genus of two species endemic to the Western Cape.

Slender to fairly robust terrestrial herbs. Leaves cauline, narrow, the lower spreading and the upper erect. Inflorescence terminal, 3–25-flowered. Flowers non-resupinate. Sepals and petals similar and free. Lip slender, flat, entire or with minute side lobes, without a spur.

Platycoryne

A genus of 19 species endemic to Africa, with only one species occurring in South Africa.

Slender, terrestrial herbs. Leaves basally tufted or on the stem. Inflorescence terminal, often a fairly dense head, few- to many-flowered. Flowers resupinate, mostly yellow, orange or greenish. Sepals free, with the median sepal forming a hood together with the petals, which are entire. Lip entire or 3-lobed; spur cylindrical or slightly club-shaped.

Platycoryne is similar to Habenaria, but distinguishable by the short, dense inflorescence, the orange-yellow flowers and by the midlobe of the rostellum, which overtops the anthers.

Platylepis

A genus of 19 species, of which nine occur in Africa, one in South Africa.

Terrestrial herbs with creeping stems rooting at the nodes; erect, flowering portion of the stem leafy. Leaves with a stalk. Inflorescence terminal and congested, many-flowered; bracts often glandular-hairy. Flowers resupinate. Sepals free. Petals partly attached to the median sepal to form a shallow hood. Lip basally fused to the column, pouch-shaped at the base.

Polystachya

A genus of 237 species distributed throughout the world's tropical regions. Its centre of diversity is in Africa, where 220 species occur, 11 of these in South Africa.

Epiphytic or rarely lithophytic, sympodial herbs, with the stems mostly forming pseudobulbs at the base. Leaves 1–6, thin-textured or leathery. Inflorescence terminal, simple or branching, 1–many-flowered; inflorescence and flower stalk often hairy. Flowers non-resupinate, occasionally hairy on the outside. Median sepal free; lateral sepals fused to each other and to the foot of the column, forming a more or less prominent chin-like extension. Lip 3-lobed, hinged at its base to the column foot, sometimes bearing a callus or hairs on its surface, without a spur.

Pterygodium

A genus of 20 species endemic to Africa, with 19 species occurring in South Africa.

Slender to robust terrestrial herbs. Leaves along the stem. Inflorescence terminal, 1–many-flowered. Flowers resupinate or not, shallowly hooded, usually strongly scented. Median sepal fused to the petals. Lip with or without side lobes, bearing an elongate appendage.

Rangaeris

A genus of five species endemic to continental Africa, with one species occurring in South Africa.

Epiphytic or lithophytic, monopodial herbs with a woody stem. Leaves leathery, in 2 opposite rows, often folded lengthwise and

bilobed at the tip. Inflorescences arising from among lower leaves, 4–15-flowered. Flowers star-shaped, small to medium-sized, white turning orange with age. Sepals and petals similar, spreading. Lip broader than the sepals and petals, with a long spur. Column short; pollinia attached to a single viscidium by individual stipes.

Rhipidoglossum

A genus of 35 species endemic to Africa, with one species occurring in South Africa.

Epiphytic, monopodial herbs with erect to hanging, woody stems. Leaves leathery or fleshy, flat to folded lengthwise, unequally bilobed at the tip. Inflorescences several to many, arising from among or below the leaves, few- to many-flowered. Flowers often semi-transparent, whitish to yellow-green. Sepals and petals similar, or petals wider and shorter than sepals, spreading. Lip spreading, usually wider than long, spurred at the base, with the tooth-like callus in the throat of the spur obscure or absent. Column stout, pollinia attached to separate viscidia by individual stipes.

Rhipidoglossum is similar to *Diaphananthe* and, until recently, the two were considered one genus; it is distinguishable by the reduced or absent tooth in the mouth of the spur and by the pollinia, which are attached to separate viscidia by individual stipes.

Satyrium

A genus of 91 species, of which 88 occur in Africa and three in Asia. There are 41 species in South Africa.

Terrestrial herbs, occasionally producing sterile shoots. Leaves usually cauline but occasionally 1 or 2 flat on the ground; sheaths on the flowering stem split to the base to fully tubular. Inflorescence few- to many-flowered, with erect, spreading or reflexed

bracts. Flowers non-resupinate. Sepals and petals similar, often basally fused or attached to the lip to form a tube. Lip forming a hood over the column, tip often extended into an erect or reflexed flap; spurs 2, elongate or pouch-shaped, rarely missing.

The former genus *Satyridium* contained a single species, which is now included in *Satyrium*.

Schizochilus

A genus of 11 species almost exclusively found in montane and subalpine grasslands of southern and south-central Africa, with eight species occurring in South Africa.

Slender, often flexuose, terrestrial herbs. Leaves linear, mostly clustered at the base. Inflorescence lax to dense, mostly nodding, few- to many-flowered. Flowers small to minute, white, yellow, mauve or a combination of these colours. Sepals and petals similar but petals smaller than sepals. Lip as long as the sepals, spurred, 3-lobed with the side lobes smaller than the midlobe.

Stenoglottis

A genus of four species endemic to continental Africa, with all four occurring in South Africa, its centre of diversity.

Lithophytic and occasionally epiphytic or terrestrial herbs. Leaves in a basal rosette, variously dotted with brown or purple. Inflorescence terminal, erect, lax to dense, few- to many-flowered. Flowers resupinate, white to mauve, often darkly spotted. Sepals and petals similar, but petals slightly shorter and wider. Lip 3–5-lobed, sometimes spurred.

Due to the substantial variation that occurs within some species, several forms have been proposed as new species and, as a result, the genus requires taxonomic revision.

Tridactyle

A genus of 47 species endemic to Africa, with four species occurring in South Africa.

Epiphytic or lithophytic, monopodial herbs forming untidy clumps; stems simple or branching, often lax and straggly. Leaves in 2 opposite rows, alternate, fleshy or leathery, unequally bilobed at the tip. Inflorescences simple, arising from among or below the lower leaves, 2–many-flowered. Flowers in 2 rows, cream to yellow or green. Sepals and petals similar, elongated, pointed, but petals slightly narrower. Lip usually 3-lobed; spur cylindrical, slender, entrance often flanked by a pair of calli.

Vanilla

A genus of 108 species distributed throughout the world's tropical regions. There are 24 species in Africa, with one species occurring in South Africa. The Central American *Vanilla planifolia* is widely cultivated for the natural flavouring extracted from its fermented fruits.

Climbing lianas with roots at the stem nodes; stems succulent, green, leafy or leafless. Leaves fleshy, elliptic, sessile or with a short stalk. Inflorescence lateral or sometimes terminal, few- to many-flowered. Flowers large, often bell-shaped. Sepals and petals free, petals often with a median keel on the outside. Lip decorated with calli or hairs. Pods long and narrow.

Ypsilopus

A genus of five species, endemic to Africa, with one species occurring in South Africa.

Epiphytic, monopodial, erect or hanging herbs with short, sometimes compressed, woody stems. Leaves narrow, usually grass-like, stiff, arranged in a fan. Inflorescences slender, arising from below the leaves, arching to hanging, 1–12-flowered. Flowers white, small and star-shaped. Sepals and petals similar, reflexed. Lip without a callus, flat, obscurely 3-lobed, with a long, slender spur from the base. Column erect, stout; pollinia 2, joined by a slender, Y-shaped stipe to a single, small viscidium.

Zeuxine

A genus of 74 species found in tropical and subtropical Africa, Asia and Australasia. There are seven species in Africa, with one in South Africa.

Terrestrial herbs with an erect, leafy stem. Leaves with or without a stalk, linear to egg-shaped. Inflorescence terminal, lax or dense, few- to many-flowered. Flowers non-resupinate, scarcely opening. Sepals and petals similar, petals converging with the median sepal to form a hood. Lip fiddle-shaped, with a pouch and 2 calli at the base.

Acampe pachyglossa

Syn: *Acampe praemorsa*

Kaapmuiden 19.01.2012

Kaapmuiden 06.02.2012

Kaapmuiden 06.02.2012

Kaapmuiden 25.02.2013

Robust, often clump-forming epiphyte with stout roots 6–8mm in diameter and somewhat woody stems, up to 300 × 10mm. **Leaves** in 2 opposite rows, succulent, unequally bilobed, up to 120–200 × 15–30mm. **Inflorescence** lateral, branched, compact, up to 150mm long, 15–25-flowered. **Flowers** fleshy, sepals and petals yellow barred with red, lip white with a few reddish spots; 15–20mm in diameter. Sepals 9–12 × 5–6mm; petals 8–10 × 3–4mm; lip 3-lobed, irregularly thickened, 9–11 × 5mm, with a sac at the base; column stout, 2mm tall.

A S O N D J F M A M J J

Restricted to a few hot and humid localities, but often locally abundant, usually along streams, from near sea level to 700m. Flowers between November and March.

Acrolophia bolusii

Cape Peninsula 24.10.2006

Cape Peninsula 08.11.2007

Cape Peninsula 28.11.2007

Robust terrestrial 300–800mm tall. **Leaves** overlapping, linear, up to 300 × 15mm. **Inflorescence** branched, laxly many-flowered. **Flowers** non-resupinate, petals and sepals brownish, lip dull yellow or white, margins occasionally suffused with mauve; 10mm wide. Sepals and petals spreading, similar, 6–7 × 2.5–3mm; lip broadly egg-shaped, often recurved, 6–9 × 6–7mm; lip margins upcurved and notched with regular, rounded teeth; disc with 10–20 slender, tall papillae; spur conical, 1mm long; column below stigma stout, 3mm tall.

Distinguishable from *A. cochlearis* by the absence of a basal constriction on the lip, shorter spur and longer sepals. The two species have little geographical overlap.

| A | S | O | N | D | J | F | M | A | M | J | J |

Found on coastal sandy flats or rarely on gravelly hills, from near sea level to 50m, but occasionally up to 450m. Flowers between October and December.

Acrolophia capensis

Cape Peninsula 30.11.2008

Still Bay 21.12.2012

Oudtshoorn 17.12.2013

Cape Peninsula 30.11.2008

Robust terrestrial 150–800mm tall. **Leaves** keeled, linear, up to 450 × 15mm.
Inflorescence simple or more commonly with up to 10 branches, laxly few- to
many-flowered. **Flowers** rather variable, sepals and petals green to purplish-
brown, lip white with purple callus and side lobes; 10–15mm long. Sepals
spreading, 9–13 × 2.5–3mm; petals slightly wider, alongside the column, with
apices reflexed; lip 3-lobed; disc with 5–7 rows of papillae; spur almost club-
shaped, 1.5–2.5mm long; column below stigma slender and laterally flattened,
5–6mm long; anther cap with 2 prominent lateral horns.

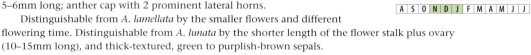

A S O N D J F M A M J J

 Distinguishable from *A. lamellata* by the smaller flowers and different
flowering time. Distinguishable from *A. lunata* by the shorter length of the flower stalk plus ovary
(10–15mm long), and thick-textured, green to purplish-brown sepals.

Widespread in fynbos, from near sea level to 2,000m. Flowers between November and January.

Acrolophia cochlearis

Mossel Bay 03.11.2007

KZN south coast 31.08.2013

Mossel Bay 01.12.2013

Robust terrestrial up to 1m tall. **Leaves** overlapping, keeled, linear, up to 400 × 15mm. **Inflorescence** with up to 5 branches, laxly to rarely densely many-flowered. **Flowers** non-resupinate, sepals and petals brownish; lip dull yellow; 7mm long. Sepals spreading, similar, 5–6mm long; median sepal slightly shorter; lip very obscurely 3-lobed, basally constricted, deeply concave with regular, round-toothed margin; disc more or less densely papillate with 3–5 rows of papillae; spur pouch-shaped, 1.5mm long; column below stigma stout, 2mm tall.

Distinguishable from *A. bolusii* by the presence of a basal constriction on the lip, longer spur and shorter sepals. The two species have little geographic overlap. Distinguishable from *A. micrantha* by the absence of side lobes on the lip and by the non-resupinate flowers.

A	S	O	N	D	J	F	M	A	M	J	J

Widespread in sand in coastal fynbos and scrub, from near sea level to 850m. Pollinated by plasterer bees. Flowers between late August and December.

Acrolophia lamellata

Cape Peninsula 28.10.2006   Cape Peninsula 08.11.2007   Cape Peninsula 29.10.2008

Cape Peninsula 28.10.2007   Cape Peninsula 30.09.2010

Robust terrestrial 150–830mm tall. **Leaves** often in a basal fan, keeled, linear to spear-shaped, margins finely toothed, up to 200 × 15mm. **Inflorescence** usually simple, occasionally with several branches; laxly many-flowered. **Flowers** with sepals and petals green to purplish-brown, lip midlobe white, side lobes purple; 7mm long. Sepals equal, spreading, 13–20 × 3–4mm; petals slightly wider, alongside the column, apices reflexed; lip 3-lobed; disc with 5–7 rows of papillae; midlobe somewhat recurved with margins notched with small, regular teeth; spur slender to club-shaped, 3–5mm long; column below stigma slender and laterally flattened, 8mm long; anther cap with 2 prominent lateral horns.

A S O N D J F M A M J J

Distinguishable from *A. capensis* by the larger flowers and different flowering time. Distinguishable from *A. lunata* by the shorter length of flower stalk plus ovary (10–15mm long), and thick-textured, green to purplish-brown sepals.

Found in coastal sandy areas, from near sea level to 400m. Flowers mainly in October, but up to December.

Acrolophia lunata

Syn: *Acrolophia barbata*

Plettenberg Bay 04.11.2008

Misgund 14.12.2011

Mossel Bay 25.11.2013

Mossel Bay 25.11.2013

Robust terrestrial up to 1.3m tall. **Leaves** often arranged in a basal fan, keeled, linear, margins finely toothed, up to 300 × 15mm. **Inflorescence** simple, rather densely many-flowered. **Flowers** white or pale rose, 12mm long. Sepals spreading, similar, 10–13 × 2.5–3mm; petals similar but slightly larger, alongside the column, apices reflexed; lip 3-lobed; disc with 5 rows of papillae; lip midlobe margins notched with small, regular teeth; spur almost club-shaped, 3mm long; column slender below the stigma, 6–7mm long.

Distinguishable from *A. capensis* and *A. lamellata* by the longer length of the flower stalk plus ovary (approximately 20mm long), and by the thin-textured, white to pale rose sepals.

A S O N D J F M A M J J

Rare in moist fynbos, from sea level to 1,700m. Flowers between November and December, with a flowering peak in the third and fourth season after fire.

Acrolophia micrantha

Betty's Bay 19.10.2013

Betty's Bay 09.10.2011

Betty's Bay 09.10.2008

Robust terrestrial 250–600mm tall. **Leaves** overlapping, keeled, linear, margins smooth or finely toothed, up to 300 × 145mm. **Inflorescence** with up to 5 branches, laxly many-flowered. **Flowers** with sepals and petals purplish-green or brown, lip white, lip side lobes purplish; 7mm long. Sepals and petals similar, 4–5 × 1.5–3mm; lip distinctly 3-lobed; disc with 3 rows of erect papillae; lip midlobe margins notched with small, regular teeth; spur pouch-shaped, 1.5mm long; column stout below the stigma, 1–2mm tall.

Distinguishable from *A. cochlearis* by the distinct side lobes of the lip and by the resupinate flowers.

A S O N D J F M A M J J

Mostly near the coast, from near sea level to 350m. Flowers between September and December.

Acrolophia ustulata

Mossel Bay 26.11.2012

Mossel Bay 25.11.2013

Mossel Bay 06.12.2003

Mossel Bay 25.11.2013

Slender terrestrial 20–100mm tall. **Leaves** overlapping, almost folded together, linear to spear-shaped, up to 50 × 5mm. **Inflorescence** simple, 3–10-flowered. **Flowers** uniformly coloured, dark maroon or rarely greenish-yellow, 8mm long. Sepals and petals similar, 7–8.5 × 2.5–3mm; lip obscurely 3-lobed; lip midlobe reflexed, densely and coarsely covered with nipple- or wart-like projections, except along the margins; lip side lobes point upwards; column slender, 6mm long.

A S O N D J F M A M J J

Very rare in sand in fynbos, from near sea level to 750m. Flowers between November and December, with a peak in the second year after fire.

Aerangis kirkii

KZN north coast 07.03.2014

KZN north coast 07.03.2014

KZN north coast 02.10.2012

Slender epiphyte with woody stem 10–50mm long and roots 1–2mm in diameter. **Leaves** 2–7, dark to greyish-green, spear-shaped with apex widest, unequally bilobed, up to 150 × 30mm. **Inflorescences** lateral, 1 to several, up to 170mm long, 2–6-flowered. **Flowers** spreading, white with a pink-tinged spur. Sepals and petals similar, 16–28mm long and 5–7mm wide; lip oblong, 16–20 × 7–8mm, with apex tapering to a long tip; spur thread-like, hanging, 60–75mm long.

A S O N D **J F M A M** J J

Very rare in coastal bush and riverine forest, from near sea level to 200m. Flowers between January and May.

Aerangis mystacidii

Eshowe 15.02.2013

Eshowe 25.02.2011

KZN north coast 11.04.2013

Eshowe 15.02.2011

Fairly robust epiphyte with woody stems up to 30mm long. **Leaves** spear-shaped, unequally bilobed, lobes rounded to tapering, 30–150 × 12–25mm. **Inflorescences** lateral, 1 to several, horizontal or hanging, 100–200mm long, 4–15-flowered. **Flowers** white. Sepals and petals similar, spreading to reflexed, 6–13 × 2–5mm; median sepal arched over the column; lip oblong-elliptic, tapering to a pointed apex, deflexed, 7–12 × 3–6mm; spur slender, 50–80mm long.

Distinguishable from *A. somalensis* by the shorter spur.

A S O N D J F M A M J J

Fairly common in subtropical coastal and submontane forests, from sea level to 800m. Flowers between February and April.

Aerangis **45**

Aerangis somalensis

Tzaneen 12.03.2013

Tzaneen 09.03.2013

Tzaneen 18.11.2012   Tzaneen 09.03.2013   Tzaneen 09.03.2013

Fairly robust epiphyte with woody stem up to 15mm long. **Leaves** 2–6, spear-shaped, often with wavy margins, deeply and almost equally bilobed, lobes blunt to tapering, grey-green with a raised darker reticulation on the upper surface, 45–110 × 20–30mm. **Inflorescences** lateral, 1 to several, 100–200mm long, 4–17-flowered. **Flowers** white. Sepals and petals similar, reflexed, 8–14 × 3–7mm; median sepal arching over the column; lip narrowly oblong, rounded to tapering, 9–15 × 4–7mm; spur slender, 100–120mm long.

Distinguishable from *A. mystacidii* by the longer spur.

| A | S | O | N | D | J | F | M | A | M | J | J |

Rare in riverine forest, from 600–1,200m. Flowers between February and May.

Aerangis verdickii

Tzaneen 03.01.2014

Tzaneen 03.01.2014

Tzaneen 03.01.2014

Tzaneen 03.01.2014

Robust epiphyte with woody stems up to 40mm long. **Leaves** strap-shaped, widest at distal end and narrowing towards the base, unequally bilobed, lobes rounded to tapering, margins wavy, 50–200 × 20–50mm. **Inflorescences** lateral, 1 to several, 200mm long, 4–12-flowered. **Flowers** white with an ivory to cream spur. Sepals unequal, median erect, arching over the column, 11–20 × 4–8mm, laterals deflexed, twisted to face backwards, 16–21 × 3–5mm; petals reflexed, 14–20 × 6–8mm; lip deflexed, 16–18 × 6–9mm; spur 120–160mm long.

Var. *verdickii* is widespread south of Rwanda and is the only variety occurring in South Africa.

Very local in rather dry woodland and montane forests, from 600–900m. Flowers between December and January.

Angraecum chamaeanthus

Kaapsehoop 06.07.2013

Graskop 12.08.2012

Kaapsehoop 06.07.2013

Kaapsehoop 06.07.2013

Minute twig epiphyte with stems less than 10mm long. **Leaves** 3–6, strap-shaped to elliptic, rather fleshy, 9–22 × 5mm. **Inflorescences** lateral, 1 to several, up to 20mm long, 5–12-flowered. **Flowers** white. Sepals 1.5 × 1mm; petals and lip slightly smaller; spur conical to spherical, 1.5mm long.

A S O N D J F M A M J J

Localized and often abundant, but mostly overlooked, in montane forests, from 1,300–1,700m. Flowers between June and August.

Angraecum conchiferum

Graskop 20.09.2010

Graskop 20.09.2010

Graskop 04.07.2013

Graskop 10.09.2013

Slender, hanging epiphyte with stems up to 300mm long, branching and often tangled. **Leaves** 8–16, adjacent, narrowly strap-shaped, unequally bilobed, 30–60 × 4–8mm. **Inflorescences** 1 to several, lateral, opposite leaves; flower stalk 20–30mm long, slender, single-flowered. **Flower** non-resupinate, sepals, petals and spur cream to yellowish-green, lip white. Sepals and petals similar, spreading, 25–30 × 2–4mm; lip very broadly egg-shaped, concave, shortly tapering to a long tip or with apiculus up to 7mm long, 12 × 10mm; spur tapering, 30–45mm long.

| A | S | O | N | D | J | F | M | A | M | J | J |

Rare, but may be locally common in cool, moist forests, often on yellowwood trees, from 200–1,600m. Flowers between September and November.

Angraecum cultriforme

KZN north coast 16.01.2013

KZN north coast 29.01.2013

KZN north coast 22.03.2013

KZN north coast 29.01.2013

KZN north coast 11.02.2013

Slender epiphyte with stems up to 250mm long. **Leaves** in 2 rows, linear to strap-shaped, very unequally bilobed with each of the lobes tapering to a point, dull green, often suffused with yellow-bronze, 50–60 × 5–8mm. **Inflorescences** 1 to several, lateral, opposite leaves, 1–4-flowered. **Flowers** pale salmon. Sepals and petals similar, somewhat reflexed, 6.5–10 × 2.5–3mm; lip deeply concave with margins and apiculus reflexed, 7 × 3mm; spur club-shaped, 14–15mm long.

A S O N D J F M A M J J

Localized, in deep shade in coastal forest, often near the ground, from near sea level to 250m. Flowers between September and February.

Angraecum pusillum

Umtamvuna 10.06.2013

Umtamvuna 26.06.2013

Umtamvuna 27.06.2013

Wilderness 30.08.2013

Umtamvuna 16.11.2011

Wilderness 30.08.2013

Dwarf epiphyte with erect stems up to 25mm long. **Leaves** 5–10, grass-like, linear, rounded, thin, in a dense terminal cluster on the stem, 40–160 × 3–4mm. **Inflorescences** several, lateral, emerging below leaves, up to 100mm long, laxly 4–15-flowered. **Flowers** white to cream. Sepals and petals similar, 1.2–1.5 × 1mm, with apices reflexed; lip deeply hooded, 2.3 × 1.5mm, 1.5mm deep; spur pouch-shaped, 1mm long.

Distinguishable from *A. sacciferum* by the white to cream flowers and inflorescence with up to 15 flowers.

A S O N D J F M A M J J

Scattered and localized, often overlooked due to its small size, growing in temperate forests, from near sea level to 1,500m. Flowers at different times in various localities, according to the timing of rainfall.

Angraecum sacciferum

Heidelberg (Western Cape) 17.01.2011

Graskop 13.12.2012

Heidelberg (Western Cape) 03.03.2011

Heidelberg (Western Cape) 03.03.2011

Dwarf twig epiphyte with erect stems up to 40mm long. **Leaves** 4–8, stiff, nearly erect, flat, linear to strap-shaped, obscurely and unevenly bilobed, 20–60 × 3–7mm. **Inflorescences** 1 to several, lateral, emerging below leaves, 1–5-flowered. **Flowers** lime-green. Sepals spreading, 3–4 × 1.5mm, with apices slightly reflexed; petals flat, 2–3 × 1mm; lip deeply hooded, 2–3mm × 1mm, 2–2.5mm deep; spur pouch-shaped, rounded, sometimes club-shaped, 2mm long.

Distinguishable from *A. pusillum* by the lime-green flowers and inflorescence with up to 5 flowers.

| A | S | O | N | D | J | F | M | A | M | J | J |

Localized, but often abundant in cool, moist forests, from near sea level to 1,800m. Flowers between November and March.

Angraecum stella-africae

Syn: *Angraecum* sp. aff. *rutenbergianum*

Vumba (Zimbabwe) 15.01.2004

Tzaneen 25.01.1976

Vumba (Zimbabwe) 15.01.2004

Tzaneen 23.02.2014

Slender epiphyte with erect stems 10–15mm long. **Leaves** 5 or 6, spreading horizontally, greenish-grey, linear, 30–50 × 6–7mm. **Inflorescences** 1 or 2, arising below the leaves, single-flowered. **Flower** greenish-white. Sepals somewhat reflexed, 15–18 × 4–5mm; petals similar to sepals, but narrower and shorter, 10–15 × 2–3mm; lip egg-shaped, flat, 13–20 × 12–14mm; spur 120–150mm long.

| A | S | O | N | D | J | F | M | A | M | J | J |

Very rare in forests, grows on *Englerophytum magalismontanum*, from 1,300–1,500m. Flowers in January.

Ansellia africana Syn: *Ansellia africana* var. *australis; Ansellia gigantea* var. *gigantea; Ansellia gigantea* var. *nilotica*

Ingwavuma 16.01.2013

Nelspruit 11.08.2012

Nelspruit 11.08.2012

Very robust epiphyte forming large clumps high up in trees; roots ascending, forming a basket at the base of the plant; pseudobulbous stems cane-like, 250–1,000 × 10–15mm. **Leaves** 4–8, strap-shaped to elliptic, pleated, leathery, with 3–7 prominent veins, shortly sheathing at the base, old sheaths persisting on the stem, 150–350 × 15–35mm. **Inflorescence** usually terminal, sometimes lateral, branched, 200–400mm long, laxly 30–50-flowered. **Flowers** yellow or greenish, uniformly coloured or variously dotted or marked with brown or maroon, 20–50mm in diameter. Sepals and petals similar, strap-shaped, 17–32 × 5–9mm; lip 3-lobed, 13–22 × 11–16mm, callus of 2–3 longitudinal ridges; column 11–15mm long.

| A | S | O | N | D | J | F | M | A | M | J | J |

Occasional in savanna or scrub forest, from 150–900m. Flowers between August and November.

Bartholina burmanniana

Op-die-Berg 18.09.2011

Op-die-Berg 25.09.2012

Op-die-Berg 14.09.2011

Cape Town 13.10.2008

Slender terrestrial up to 230mm tall. **Leaf** 1, flat on the ground, circular, 10–40 × 8–20mm. **Inflorescence** hairy, single-flowered. **Flower** with sepals greenish, petals and lip greyish-white to pale mauve. Sepals hairy, slightly recurved, 6–12mm long; petals linear to spear-shaped, recurved, 8–14mm long; lip spreading, fan-shaped, up to 35 × 30mm, divided into 4–6 lobes with up to 26 thread-like segments; spur 8–12mm long.

| A | S | O | N | D | J | F | M | A | M | J | J |

Occasional in fynbos or restioveld, from near sea level to 2,000m. Flowers between August and October, stimulated by fire.

Bartholina etheliae

Barrydale 12.10.2012

Ceres 15.11.2009

Ceres 16.11.2008

George 17.10.2011

Slender terrestrial up to 300mm tall. **Leaf** 1, flat on the ground, circular, 10–36 × 8–24mm. **Inflorescence** hairy, single-flowered. **Flower** with sepals green, petals and lip pale lilac-blue, terminal knobs white. Sepals 10–20mm long; petals linear to spear-shaped, 8–20mm long; lip spreading, fan-shaped, up to 40 × 35mm, divided into 4–6 lobes comprised of numerous thread-like segments with terminal knobs; spur 8–14mm long.

A S O N D J F M A M J J

Rare or localized in dry fynbos and renosterveld, often growing under bushes, from near sea level to 1,800m. Flowers between October and December, and as early as July in the more arid areas.

Bolusiella maudiae

Eshowe 30.01.2013

Eshowe 25.02.2011

Eshowe 09.02.2014

Eshowe 08.02.2013

Miniature epiphyte up to 50mm tall, stemless, roots thin and numerous. **Leaves** 3–10, succulent, laterally compressed and overlapping to form a fan. **Inflorescence** lateral, ascending, 8–16-flowered. **Flowers** in 2 nearly opposite rows, white with a green spur; 4mm wide. Sepals and petals similar, 4 × 1mm; lip 2.5 × 1mm; spur rounded or slightly inflated near the tip, 1.5mm long.

A S O N D J F M A M J J

Localized, but often abundant in moist lowland forests, from 100–600m. Flowers between January and February.

Bonatea antennifera

Syn: *Bonatea speciosa* var. *antennifera*

Johannesburg 23.03.2013

Jozini 08.06.2013

Johannesburg 12.03.2012

Johannesburg 23.03.2013

Robust terrestrial up to 1.2m tall. **Leaves** 6–21, cauline, oblong to broadly spear-shaped, up to 190 × 53mm. **Inflorescence** lax to dense, 4–44-flowered. **Flowers** green and white. Median sepal erect, 14–23mm long; lateral sepals deflexed, 13–26mm long; petals divided; upper petal lobe linear, 13–24mm; lower petal lobe narrowly linear throughout, projecting forward and ascending, 24–43 × 0.5–2mm; lip 3-lobed from a short, narrow base; lip midlobe narrowly linear, sharply bent near the middle, 15–34mm long; lip side lobes narrowly linear, with apex recurved, 17–45 × 0.5–2mm; spur somewhat swollen towards apex, 27–44mm long.

A S O N D J F M A M J J

    Distinguishable from *B. boltonii* by the forward-projecting, ascending lower petal lobes and the 15–22mm-long stigmatic arms with forward-projecting, club-shaped stigma. Distinguishable from *B. speciosa* by the shorter (less than 15mm), club-shaped apex of the spur and lower petal lobe less than 2mm wide.

Occasional in grassland, savanna and dry woodland, from near sea level to 1,700m. Flowers between March and June.

Bonatea boltonii

Nelshoogte Pass 17.02.2010

KZN south coast 15.08.2012

KZN south coast 30.09.2014

KZN south coast 30.09.2014

Slender terrestrial of variable height, 100–750mm tall. **Leaves** 5–13, cauline, oblong to broadly spear-shaped, up to 178 × 40mm. **Inflorescence** lax to dense, 3–20-flowered. **Flowers** green and white. Median sepal erect, 12–19mm long; lateral sepals slightly deflexed, 10–20mm long; petals divided; upper petal lobe 9–18mm; lower petal lobe linear, spreading horizontally, 11–24 × 0.5–2mm; lip 3-lobed from a short, narrow base; lip midlobe narrowly linear, sharply bent near the middle, 7–27mm long; lip side lobes linear to spear-shaped, descending, diverging, 10–32 × 0.5–3mm; spur somewhat swollen towards apex, 27–51mm long.

Distinguishable from *B. antennifera* by the shorter stigmatic arms with a spatula-shaped, deflexed stigma. Distinguishable from *B. speciosa* by a shorter (less than 15mm), club-shaped spur apex.

Rare in grassland, savanna and dunes, from near sea level to 1,900m. Flowers between August and September along the coast, and between December and March elsewhere.

Bonatea cassidea

Syn: *Bonatea saundersiae*

Port Alfred 02.09.2012

Port Alfred 30.09.2011

Eshowe 28.07.2012

Port Alfred 02.09.2011

Eshowe 28.07.2012

Slender terrestrial up to 640mm tall. **Leaves** 5–13, cauline, linear to spear-shaped, sometimes withered at flowering time, up to 250 × 19mm. **Inflorescence** lax, 3–35-flowered. **Flowers** green and white, lip midlobe and upper petal lobes white in western populations and green in eastern populations. Median sepal erect to almost erect, 7–13.5mm long; lateral sepals strongly deflexed, 7–15.5mm long; petals divided; upper petal lobe erect, narrowly linear, 6–14mm long; lower petal lobe spear-shaped to almost egg-shaped, spreading horizontally, 8.5–15 × 3–8mm; lip 3-lobed from a narrow base; lip midlobe narrowly linear, bent, 5.5–17mm long; lip side lobes wide and spreading, 6.5–21 × 2–6.5mm; spur 12–26mm long.

A	S	O	N	D	J	F	M	A	M	J	J

Fairly common in savanna, forest and thicket, from near sea level to 1,500m. Pollinated by butterflies. Flowers between July and October.

Bonatea lamprophylla

KZN north coast (cultivated plant) 22.10.2012

KZN north coast (cultivated plant) 22.10.2012

KZN north coast 24.10.2013

KZN north coast (cultivated plant) 22.10.2012

Robust terrestrial up to 1m tall. **Leaves** 5–15, cauline, egg-shaped, margin crisped, up to 130 × 75mm. **Inflorescence** lax, 5–16-flowered. **Flowers** green and white. Median sepal erect, 28.5–34mm long; lateral sepals strongly deflexed, 28.5–33.5mm long; petals divided; upper petal lobe linear, 27.5–32.5mm long; lower petal lobe linear, spreading horizontally and curved upward, 34.5–54 × 0.5–1.5mm; lip 3-lobed from a narrow base; lip midlobe linear, sharply bent near the middle; lip side lobes thread-like, descending and diverging, 106–135 × 0.5–1mm; spur somewhat swollen in apical half, rather flattened, 92–131mm long.

Distinguishable from *B. steudneri* by the longer lip side lobes.

| A | S | O | N | D | J | F | M | A | M | J | J |

Rare in deep shade in coastal dune forest, from near sea level to 170m. Flowers between September and October.

Bonatea polypodantha

Magaliesberg 21.02.2014

Muden 05.03.2013

Stutterheim 28.02.2011

Slender terrestrial up to 330mm tall. **Leaves** 4–8; basal leaves 1–3, broadly elliptic, up to 120 × 35mm; **cauline leaves** 2–5, narrowly spear-shaped, much smaller than basal leaves. **Inflorescence** lax, 2–12-flowered. **Flowers** pale green and white. Median sepal erect, 7–11mm long; lateral sepals spreading, 6–10 × 3.5–7mm; petals divided; upper petal lobe erect, linear, 7–12mm long; lower petal lobe thread-like, spreading horizontally, variously curved, 22.5–35.5 × 0.5–1mm; lip 3-lobed with lobes thread-like from a short undivided base; lip midlobe 11–22mm long; lip side lobes descending, strongly divergent, 25–48 × 0.5–1.5mm; spur slightly swollen, 31–47mm long.

| A | S | O | N | D | J | F | M | A | M | J | J |

Distinguishable from *B. pulchella* by the smaller flowers, shorter spur and difference in flowering time.

Fairly common in riverine scrub forest in dense shade, from 550–2,000m. Flowers between January and April.

Bonatea porrecta

Peacevale 21.08.2011

Peacevale 21.08.2011

Peacevale 23.08.2011

Fairly robust terrestrial up to 560mm tall. **Leaves** 5–16, cauline, narrowly oblong, withered at flowering time, up to 131 × 25mm. **Inflorescence** lax, 3–36-flowered. **Flowers** green and white. Median sepal erect, with apex reflexed, 8.5–15mm long; lateral sepals reflexed, 8–14.4mm long; petals divided; upper petal lobe linear, 6.5–16mm long; lower petal lobe narrowly linear, spreading horizontally, curved towards apex, 12.5–24.5 × 0.5–1.5mm; lip 3-lobed from a short, narrow base; lip midlobe narrowly linear, shorter than side lobes, bent near the middle, 10.5–20mm long; lip side lobes narrowly linear, curled near apex, 14–29mm long; spur swollen towards apex, 24–43mm long.

| A | S | O | N | D | J | F | M | A | M | J | J |

Occasional in grassland and scrub, from 150–1,800m. Flowers between June and August.

Bonatea pulchella

Alkmaar 11.05.2012

Alkmaar 11.05.2012

Nelspruit 10.05.2013

Nelspruit 10.05.2013

Slender terrestrial up to 320mm tall. **Leaves** 5–9; basal leaves in a rosette of 2–5, broadly oblong to elliptic, 27–113 × 12–44mm; **cauline leaves** few, narrowly spear-shaped, 13–39 × 2–9mm. **Inflorescence** lax, 2–11-flowered. **Flowers** white, back of sepals and spur apices green. Median sepal erect, 10–15.5mm long; lateral sepals 9.5–14.5mm long; petals divided; upper petal lobe narrowly linear, 9–15mm long; lower petal lobe thread-like, spreading, 36–50 × 0.5–1mm; lip 3-lobed from a narrow base; lip midlobe very narrowly linear, 21.5–38mm long; lip side lobes thread-like, 36–57.5mm long; spur swollen towards apex, 53–70mm long.

Distinguishable from *B. polypodantha* by the larger flowers, longer spur and difference in flowering time.

A	S	O	N	D	J	F	M	A	M	J	J

Occasional in coastal forest and thicket, growing in shade, from near sea level to 800m. Flowers between May and July.

Bonatea saundersioides

Magaliesberg 07.04.2013

Magaliesberg 07.04.2013

Magaliesberg 07.04.2013

Magaliesberg 07.04.2013

Magaliesberg 07.04.2013

Slender terrestrial up to 600mm tall. **Leaves** 5–12, cauline, narrowly elliptic, may have started withering at flowering time, up to 193 × 28mm. **Inflorescence** lax, 4–41-flowered. **Flowers** green and white. Median sepal erect, 8.5–15mm long; lateral sepals strongly deflexed, 10–15mm long; petals divided; upper petal lobe linear, 8–14mm long; lower petal lobe spreading, curved towards apex, 16–24 × 0.5–2mm; lip 3-lobed from a narrow base; lip midlobe narrowly linear, tapering, bent near the middle, 8.5–20mm long; lip side lobes narrowly linear, curved, 14–34 × 0.5–2mm; spur curved, 16–30mm long.

A S O N D J F M A M J J

Scattered in shade in wooded ravines and savanna, from 600–1,400m. Flowers between April and July.

Bonatea speciosa

Gansbaai 12.10.2010

Wilderness 21.10.2010

Hibberdene 28.06.2013

Wilderness 22.08.2012

Hibberdene 28.06.2013

Robust terrestrial up to 950mm tall. **Leaves** 4–19, cauline, oblong to broadly spear-shaped, up to 155 × 67mm. **Inflorescence** rather dense, 1–41-flowered. **Flowers** green and white. Median sepal nearly erect to erect, 5–25mm long; lateral sepals spreading, 14–29.5mm long; petals divided; upper petal lobe linear, tapering, 13–26mm long; lower petal lobe linear to spear-shaped, projecting forward, 19–39 × 2–5.5mm; lip 3-lobed from a short, narrow base; lip midlobe linear, sharply bent near middle, 9–27.5mm long; lip side lobes narrowly linear, descending and diverging with recurved apex, 11.5–39 × 1–4mm; spur slightly swollen towards apex, 15–32mm long.

Distinguishable from *B. antennifera* and *B. boltonii* by the longer (more than 15mm), club-shaped apex of the spur, and the lower petal lobe more than 2mm wide, projecting forward and deflexed.

A S O N D J F M A M J J

Common in sandy soils in coastal scrub, on forest margins, and sometimes in savanna, from near sea level to 1,300m. Pollinated by hawkmoths. Flowers between June and December.

Bonatea steudneri

Kitengela (Kenya) 17.05.2005

Robust terrestrial up to 1m tall. **Leaves** 7–20, cauline, elliptic to spear-shaped, up to 154 × 56mm. **Inflorescence** lax, 5–35-flowered. **Flowers** green and white. Median sepal erect, 19–25mm long; lateral sepals reflexed, 17.5–27.5mm long; petals divided; upper petal lobe narrowly linear, 18–26 × 1mm; lower petal lobe narrowly linear to thread-like, 39–75 × 0.5–2mm; lip 3-lobed from a narrow base; lip midlobe descending, bent, 22–43mm long; lip side lobes narrowly linear, descending and diverging, 35–60 × 0.5–2mm; spur slightly swollen towards apex, 75–230mm long.

Distinguishable from *B. lamprophylla* by the shorter lip side lobes.

| A | S | O | N | D | J | F | M | A | M | J | J |

Very rare in thicket, grassland and savanna, from near sea level to 700m. Pollinated by hawkmoths. Flowers between March and May.

Brachycorythis conica

Krugersdorp 05.02.2012

Krugersdorp 12.01.2013

Krugersdorp 12.01.2013

Krugersdorp 05.02.2012

Krugersdorp 05.02.2012

Slender to fairly robust terrestrial up to 400mm tall. **Leaves** numerous, cauline, densely overlapping, spear-shaped to narrowly egg-shaped, up to 60 × 20mm. **Inflorescence** dense, up to 150mm long, 30–40-flowered. **Flowers** mauve to pink or white, lip mauve with purple spots on upper surface; 20 × 10mm. Sepals up to 10 × 3mm; petals shielding column, up to 10 × 4mm; lip up to 12mm long, with apex blunt and square or shortly 3-lobed, up to 8 × 8mm; lip limb margins with small rounded teeth; spur curving downward, 5.5–8mm long.

Subsp. *transvaalensis* is restricted to South Africa and is the only subspecies occurring in the region.

| A | S | O | N | D | J | F | M | A | M | J | J |

Very rare in open grassland, from 800–1,700m. Flowers between January and April.

Brachycorythis inhambanensis

KZN south coast 19.12.2011

KZN south coast 19.12.2011

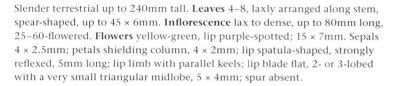

KZN south coast 19.12.2011

KZN south coast 19.12.2011

KZN south coast 19.12.2011

Slender terrestrial up to 240mm tall. **Leaves** 4–8, laxly arranged along stem, spear-shaped, up to 45 × 6mm. **Inflorescence** lax to dense, up to 80mm long, 25–60-flowered. **Flowers** yellow-green, lip purple-spotted; 15 × 7mm. Sepals 4 × 2.5mm; petals shielding column, 4 × 2mm; lip spatula-shaped, strongly reflexed, 5mm long; lip limb with parallel keels; lip blade flat, 2- or 3-lobed with a very small triangular midlobe, 5 × 4mm; spur absent.

A S O N D J F M A M J J

Very rare in damp grassland, from sea level to 400m. Flowers between December and March.

Brachycorythis macowaniana

George 02.12.2007 George 02.12.2007 Joubertina 14.11.2010

George 25.11.2013 George 25.11.2013

Slender to fairly robust terrestrial up to 320mm tall. **Leaves** 5–15, closely overlapping along stem, spear-shaped to narrowly spear-shaped, up to 60 × 17mm. **Inflorescence** lax to dense, up to 130mm long, many-flowered. **Flowers** with sepals reddish-brown, petals and lip pale green; 12 × 5mm. Sepals shielding the column, 4 × 2mm; petals 3 × 2mm; lip spatula-shaped, strongly recurved, distinctly 3-lobed apically, 4mm long; spur almost club-shaped, 1.7–2.5mm long.

A S O N D J F M A M J J

Rare in grassland or restioveld, from 150–1,500m. Flowers between November and December, after fire.

Brachycorythis ovata

Himeville 15.12.2008

Himeville 15.12.2008

Himeville 03.12.2012

Umtamvuna 01.12.2013

Himeville 22.01.2008

Robust to slender terrestrial up to 400mm tall. **Leaves** stiff, numerous, cauline, closely overlapping, spear-shaped, up to 70 × 25mm. **Inflorescence** dense, up to 270mm long, 20–100-flowered. **Flowers** rather fleshy, white with suffused shades of mauve-pink to purple, lip keel white with irregular purple spots; 20 × 10mm. Median sepal 5–6 × 3–4mm; lateral sepals 7–8 × 3–4mm; petals closely placed on each side of the column, 6–7 × 3–5mm; lip 8–16mm long; lip limb boat-shaped; lip blade strongly keeled, usually 3-lobed with midlobe longer than, equalling or shorter than side lobes; spur absent.

A S O N D J F M A M J J

Subsp. *ovata* is restricted to South Africa and Swaziland and is the only subspecies occurring in the region.

Distinguishable from *B. pubescens* by plants that are not hairy, and the absence of a yellow callus at the base of lip.

Sporadic in grassland, from near sea level to 1,800m. Flowers between September and January.

Brachycorythis pleistophylla

Brondal 03.12.2011

Bushbuckridge 20.11.2013

Bushbuckridge 20.11.2013

Bushbuckridge 20.11.2013

Robust terrestrial up to 1m tall. **Leaves** numerous, cauline, similar in size and shape, lax to densely overlapping, linear to spear-shaped, up to 60 × 15mm. **Inflorescence** moderately dense, up to 150mm long, 15–80-flowered. **Flowers** rose-pink or purple, often with a white or yellow centre, 20 × 10mm. Sepals and petals recurved; sepals 6–9 × 3mm; petals 4–6 × 3–5mm; lip spatula-shaped, 7–13 × 6–14mm; lip limb short and narrow; lip blade 2-lobed, sometimes with a short, tooth-like midlobe present; spur absent.

Var. *pleistophylla* (syn: *Brachycorythis pleistophylla* subsp. *pleistophylla*) is widespread in Africa, and is the only variety occurring in South Africa.

A S O N D J F M A M J J

Rare in montane grassland, from 1,300–1,850m. Flowers between November and December.

Brachycorythis pubescens

Giant's Castle 16.12.2008

Giant's Castle 19.01.2014

Giant's Castle 16.12.2008

Giant's Castle 16.12.2008

KZN south coast 30.11.2013

Moderately robust terrestrial up to 600mm tall. **Leaves** numerous, cauline, closely overlapping, hairy, spear-shaped to narrowly spear-shaped, up to 60 × 20mm. **Inflorescence** dense, hairy, up to 320mm long, 45–110-flowered. **Flowers** hairy on the outside, light to deep pink suffused with brownish-green and with a red-spotted yellow callus at base of the lip; 20 × 10mm. Sepals and petals forming a hood over the column; sepals 5–7 × 3–4mm; petals 4–6 × 2–3mm; lip spatula-shaped, bent from limb to blade, 7–10mm long; lip blade 3-lobed with broadly diverging side lobes; spur absent.

Distinguishable from *B. ovata* subsp. *ovata* by plants that are hairy, and a yellow callus at the base of the lip.

A	S	O	N	D	J	F	M	A	M	J	J

Sporadic in grassland, from sea level to 2,200m. Flowers between October and January.

Brachycorythis tenuior

Cato Ridge 01.02.2014   Umzinto 24.01.2009   Cato Ridge 01.02.2014

Umzinto 24.01.2009   Umzinto 24.01.2009

Slender terrestrial up to 250mm tall. **Leaves** several, cauline, lax to closely overlapping, spear-shaped, up to 38 × 13mm. **Inflorescence** lax, up to 120mm long, 6–20-flowered. **Flowers** white suffused with pink to purple, spotted purple with lip limb uniform pink to maroon-purple; 20 × 10mm. Sepals 7–10 × 3–5mm, laterals reflexed, ascending and wing-like; petals 7–8 × 4–5mm; lip spatula-shaped; lip limb short with 2 tall keels; lip blade triangular; spur stout, almost club-shaped, 5.5–10.5mm long.

A S O N D J F M A M J J

Rare and sporadic in grassland, from 300–1,650m. Flowers between January and March.

Brownleea coerulea

Stutterheim 01.03.2011

Graskop 24.03.2010

Sabie 25.02.2013

Luneburg 11.02.2014

Sabie 28.02.2013

Slender terrestrial or lithophyte 100–600mm tall. **Leaves** usually 3, alternate, narrowly egg-shaped, 80–220 × 35–40mm. **Inflorescence** lax, nearly 1-sided, 30–100mm long, 3–30-flowered. **Flowers** mauve, often with deep purple markings on galea; 20mm in diameter. Median sepal 8–10mm long; galea 1mm wide, 3mm deep; spur slender, frequently decurved, with apex almost club-shaped, 13–26mm long; lateral sepals 8–13mm long; petals exserted from galea, apex nearly truncated; lip 1mm long.

Distinguishable from *B. graminicola* by the wider, egg-shaped leaves, gently decurved spur, and forest habitat.

A S O N D J F M A M J J

Locally common in shaded habitats, usually in forests or forest patches, in moss and on rock ledges, from sea level to 1,800m. Pollinated by long-proboscid flies. Flowers between February and April.

Brownleea galpinii

Very slender terrestrial 150–500mm tall. **Leaves** usually 2, cauline, nearly folded together; lower leaf larger, prominently ribbed, 60–200 × 5–15mm. **Inflorescence** head-like, often 1-sided, 5–25-flowered. **Flowers** white to cream with small purple spots on the petals; 5mm in diameter. Median sepal hooded, 5–10mm tall, 2mm deep; spur slender, straight, 3.5–6mm long; lateral sepals 6–10 × 2–4mm; petals fiddle-shaped with apical lobe spreading fan-like; lip 0.8–2.5mm long.

Brownleea galpinii subsp. *galpinii*

Greytown 17.03.2010

Sabie 24.02.2014

Greytown 17.03.2010

Greytown 17.03.2010

Subsp. *galpinii* is distinguished as follows: lateral sepals 6–7mm long; median sepal 5–8mm long; petals with apical lobe not notched or only slightly notched with regular teeth; lip erect, 0.8–1.3mm long, not reaching to the top of the stigma.

Subsp. *galpinii* occurs only in South Africa and Zimbabwe.

| A | S | O | N | D | J | F | M | A | M | J | J |

Rare in rocky grassland, and occasionally in seepages, from 1,200–2,200m. Flowers between February and April.

Brownleea galpinii subsp. *major*

Sentinel 06.02.2007

Witsieshoek 29.02.2008

Witsieshoek 09.02.2009

Sentinel 03.03.2009

Subsp. *major* is distinguished as follows: lateral sepals 8–10mm long; median sepal 7.5–9mm long; petals with apical lobe fan-like and notched with small, regular teeth; lip 2–2.5mm long, reaching over the stigma to between rostellum lobes.

Subsp. *major* occurs only in South Africa and Lesotho.

Distinguishable from *Disa cephalotes* subsp. *cephalotes* by the apparent absence of a lip.

Occasional in grassland or in open *Protea* savanna in basaltic or sandstone-derived soil, from 1,800–2,500m. A mimic of *Scabiosa* flowers. Pollinated by long-proboscid flies. Flowers between the end of January and March.

| A | S | O | N | D | J | F | M | A | M | J | J |

Brownleea graminicola

Dullstroom 27.03.2010

Dullstroom 27.03.2010

Dullstroom 07.04.2013

Dullstroom 27.03.2010

Dullstroom 07.04.2013

Slender terrestrial 240–340mm tall. **Leaves** usually 3, cauline, spirally arranged, narrowly spear-shaped, underside with 3 obvious veins, 80–110 × 18–24mm. **Inflorescence** fairly dense, 12–25-flowered. **Flowers** bluish to lilac to dark lilac; 10mm in diameter. Median sepal 7–9 × 4mm; galea recurved apically; spur straight to very gently upcurved with mouth narrowing abruptly and apex almost club-shaped, 19–23mm long; lateral sepals 9–11 × 5–6mm; lip minute, 1 × 0.1mm.

Distinguishable from *B. coerulea* by the narrower, spear-shaped leaves, generally straight spur, and grassland habitat.

| A | S | O | N | D | J | F | M | A | M | J | J |

Rare on rocky, open, sunny sourveld grassland on the eastward side of sheltering rocks and boulders, from 1,800–2,200m. Flowers between March and April.

Brownleea macroceras

Naudesnek 28.01.2009

Naudesnek 04.03.2012

Naudesnek 07.02.2008

Naudesnek 07.02.2008

Slender terrestrial or lithophyte 100–300mm tall. **Leaves** usually 1, narrowly spear-shaped, the sides nearly folded together, 60–80 × 4–10mm. **Inflorescence** usually single-flowered, but sometimes up to 6-flowered. **Flowers** facing outward horizontally, sepals and petals pale mauve to almost white; up to 30mm in diameter. Median sepal 10–13mm long; spur slender, somewhat decurved and almost club-shaped near the tip, 25–40mm long; lateral sepals 13–18 × 4–6mm; petals facing forward; lip 2mm long.

In the northern part of the distribution range, plants tend to be smaller, with a single leaf and a single, larger flower; in the south, plants tend to be bigger, occasionally with up to 3 leaves and 6 smaller flowers.

A S O N D **J F M A** M J J

Occasional on rocky outcrops, rock ledges and in shallow rocky soil, from 1,800–2,700m. Pollinated by long-proboscid flies. Flowers between late January and the beginning of April.

Brownleea parviflora

Sani Pass 28.02.2009

Garden Castle 09.02.2008

Witsieshoek 02.03.2008

Maclear 19.02.2008

Maclear 19.02.2008

Sabie 03.03.2013

Slender terrestrial 200–600mm tall. **Leaves** 3, narrowly spear-shaped, grading into floral bracts; basal leaf the longest, 80–200 × 8–16mm. **Inflorescence** dense, 20–60-flowered. **Flowers** white with slight green or brown tint; 4mm in diameter. Median sepal 3–5mm long; spur soon sharply decurved, frequently club-shaped, 3–5mm long; lateral sepals 3–5mm long; petals oblong to almost square; lip 0.5–1mm long.

A S O N D J F M A M J J

Occasional in damp montane grassland, sometimes among rocks, from sea level to 3,000m. Pollinated by anthophorid bees. Flowers between February and March.

Brownleea recurvata

Elliot 27.02.2011

Naudesnek 22.02.2007

Naudesnek 26.02.2007

Naudesnek 04.03.2012

Naudesnek 27.02.2011

Slender terrestrial 200–500mm tall. **Leaves** 2, linear to narrowly spear-shaped; lower leaf the largest, up to 200 × 8mm. **Inflorescence** lax, 12–17-flowered. **Flowers** white to soft pink, spur dark pink and upper petal lobe with a few dark spots; 10mm in diameter. Median sepal 7–12mm long; spur occasionally club-shaped, decurved from the middle, 5–10mm long; lateral sepals 8–11mm long; petals narrowly oblong, with a small lobe at the base.

A S O N D J F M A M J J

Localized in well-drained localities, often in pebbly soils, or rarely in damp localities, from near sea level to 1,500m. Flowers between the end of February and the beginning of April, stimulated by fire.

Bulbophyllum elliotii

Tzaneen 13.03.2014

Chimanimani (Zimbabwe) 16.04.2014

Vumba (Zimbabwe) 01.03.2003

Slender epiphyte or lithophyte up to 50mm tall, with short rhizomes 1.5mm in diameter; pseudobulbs clustered, ovoid, usually flattened, 10–15mm in diameter. **Leaves** 2 per pseudobulb, fleshy, stiff, rounded, 40–50 × 10–15mm. **Inflorescence** lateral, curving downward, 10–30-flowered; rachis fleshy, rounded. **Flowers** fleshy, papillose on the outside, not opening fully, deep purple-red or greenish. Sepals 4.6 × 1.6mm; petals linear, tapering to a fine point, 2 × 0.2mm; lip fleshy, 2 × 1mm, with a fringe of long hairs along the margin and 2 bulbous fleshy ridges on underside.

A S O N D J F M A M J J

Very rare in exposed areas on the escarpment, from 900–1,650m. Flowers between February and March.

Bulbophyllum longiflorum

KZN north coast 24.10.2013

KZN north coast 24.10.2013

KZN north coast 21.03.2013

KZN north coast 24.10.2013

Slender to fairly robust epiphyte with stout rhizomes 3–4mm in diameter; pseudobulbs well spaced, narrowly conical, scarcely ribbed, 30–40 × 10–15mm. **Leaf** 1, leathery, narrowly oblong, 130–180 × 20–40mm. **Inflorescence** lateral, up to 150mm tall, with all the flowers arising from a single point, 3–6-flowered; inflorescence stalk slender. **Flowers** pinkish-yellow to pale orange-yellow, spotted with purple. Sepals unequal, median 8–10 × 7mm, with a bristle-like tip, 4.5–8mm long, laterals fused in apical two thirds, 20–25 × 4–5mm; petals 7 × 3mm, with a bristle-like tip up to 4mm long; lip entire, fleshy, 7 × 2.5mm.

A S O N D J F M A M J J

Very rare in riverine forest, from near sea level to 250m. Flowers between October and January.

Bulbophyllum sandersonii

Nkandla 10.11.2012

Nkandla 09.12.2012

Nkandla 29.11.2012

Luneburg 10.12.2012

Nkandla 29.11.2012

Slender to fairly robust epiphyte or lithophyte with long rhizomes 2–3mm in diameter; pseudobulbs ovoid to narrowly conical, 4- or 5-angled, 30–50mm apart, 20–40 × 10mm. **Leaves** 2 per pseudobulb, leathery to almost fleshy, strap-shaped to elliptic-oblong, 25–100 × 10–17mm. **Inflorescence** lateral, erect, 100–150mm tall, 6–16-flowered; rachis swollen, flattened. **Flowers** borne in 2 rows, dull violet or green with purplish dots. Sepals papillose, unequal, median 6.5–11 × 1–1.25mm, laterals 3.5–6 × 2.5–4mm, apiculus 1.5mm long; petals 5–10 × 0.4–1mm; lip fleshy, 2 × 1mm.

| A | S | O | N | D | J | F | M | A | M | J | J |

Subsp. *sandersonii* is found in our region and is widespread in Africa.

Distinguishable from *B. scaberulum* by the flat, erect median sepal, and petals with a club-shaped tip.

Occasional, but often forms large colonies in montane or temperate forests, from 800–1,600m. Flowers between October and December.

Bulbophyllum scaberulum

Eshowe 09.11.2012

Graskop 26.10.2012

Graskop 18.11.2012

Umtamvuna 16.11.2012

Slender to fairly robust epiphyte or lithophyte; pseudobulbs ovoid to narrowly conical, 4- or 5-angled, 30–50mm apart, 15–30 × 10–17mm. **Leaves** 2 per pseudobulb, leathery, narrowly oblong, 25–80 × 15–20mm. **Inflorescence** lateral, 90–260mm tall, 24–40-flowered; rachis swollen, flattened. **Flowers** open, dull purplish-violet, lateral sepals with inner surface striped white and purple. Sepals unequal, median linear, concave, hooding the flower, with apex recurved, 6–10 × 1.5–2mm; laterals 4–8 × 2.5mm; petals 2.5–6 × 0.5–1mm; lip fleshy, 2–6 × 2–3mm.

Var. *scaberulum* is found in our region and is widespread in Africa.

Distinguishable from *B. sandersonii* by the concave, upward-pointing median sepal and linear, outward-pointing petals without a club-shaped tip.

| A | S | O | N | D | J | F | M | A | M | J | J |

Occasional, but often forming large colonies in light shade and moist forests, from near sea level to 700m. Flowers between October and December.

Calanthe sylvatica

Weza 04.02.2014

Kaapsehoop 16.02.2011

Eshowe 09.02.2014

Eshowe 09.02.2014

Robust terrestrial up to 700mm tall. **Leaves** 3–5, pleated, thin, 200–400 × 70–115mm. **Inflorescence** terminal, dense to lax, hairy, 10–20-flowered. **Flowers** white to pink or pale mauve, fading to apricot. Sepals and petals similar, 12–27 × 5–12mm; lip 3-lobed with base extensively fused to the column, 9–18 × 7–15mm; lip midlobe with 2 large, rounded lobes at the apex, callus of 3 small ridges near the base; spur 15–30mm long.

A S O N D J F M A M J J

Widespread in deep shade in temperate or submontane forests, often near streams or seepages, occasionally on tree trunks, from near sea level to 1,700m. Flowers between December and March.

Centrostigma occultans

Dullstroom 04.02.2011

Dullstroom 06.02.2012

Dullstroom 28.01.2012

Dullstroom 27.01.2011

Dullstroom 31.01.2011

Robust terrestrial up to 600mm tall. **Leaves** 5–10, linear to spear-shaped, up to 140 × 20mm. **Inflorescence** lax, 3–7-flowered. **Flowers** pale yellow to greenish-yellow. Sepals unequal, median hooded, erect, with apex recurved, 16mm long; laterals spear-shaped, spreading, 17mm long; petals strap-shaped, 15mm long; lip 3-lobed from a short, undivided base; lip midlobe entire, linear, 20 × 2mm; lip side lobes spreading, fringed, 20 × 3mm; spur hanging, 120mm long.

A S O N D J F M A M J J

Very rare in grassland marshes, from 1,200–2,100m. Flowers between January and February.

Ceratandra atrata

Betty's Bay 29.11.2008

Wolseley 06.11.2007

Cape Peninsula 27.10.2006

Op-die-Berg 04.12.2010

Slender terrestrial up to 350mm tall. **Leaves** numerous, narrowly linear, up to 60 × 4mm. **Inflorescence** lax to fairly dense, up to 170mm long, many-flowered; bracts bearing a fringe of hairs along the margin. **Flowers** non-resupinate, greenish-yellow. Median sepal fused to petals, 12 × 3–4mm; lateral sepals 11–12 × 5–6mm; petals 12–13 × 5mm; lip erect, anchor-shaped, usually with fleshy callus; rostellum arms horn-like, projecting, 7–8mm long.

A S O N D J F M A M J J

Locally common in marshes or on wet rocky outcrops and stream banks, from near sea level to 1,500m. Pollinated by oil-collecting bees. Hybridizes with *C. grandiflora*. Flowers between October and December, after fire.

Ceratandra bicolor

Cape Peninsula 28.11.2007

Franschhoek 17.12.2007

Cape Peninsula 30.11.2008

Cape Peninsula 28.11.2007

Slender terrestrial up to 340mm tall. **Leaves** numerous, linear to spear-shaped, up to 40 × 2mm. **Inflorescence** lax, up to 110mm long, 1–12-flowered. **Flowers** with sepals green, petals and lip yellow. Median sepal 10–12 × 3–5mm; lateral sepals 11–13 × 5–8mm; petals 10–12 × 6–7mm; lip 6–9 × 10–12mm, with 2 broad, divergent apical lobes 2–3mm wide; lip blade with fleshy callus; lip appendage a pair of horns, 6–9mm tall; rostellum lobes divergent, 8mm long, curving around base of lip appendage.

A S O N D J F M A M J J

Rather local in fynbos, from 90–1,100m. Pollinated by oil-collecting bees. Flowers between November and December, after fire.

Ceratandra globosa

Ceres 04.12.2010

Porterville 30.11.2009

Ceres 05.12.2013

Ceres 30.11.2009

Slender terrestrial up to 400mm tall. **Leaves** numerous, narrowly linear, up to 80 × 5mm. **Inflorescence** head-like, up to 60mm long, 6–many-flowered; bracts without hairs. **Flowers** with sepals pale purplish-green to pink, petals and lip white, rostellum maroon. Median sepal fused to the petals, 5–7 × 2–3mm; lateral sepals 5–7 × 3–4mm; petals 6–7 × 4–5mm; lip spatula-shaped, 4–5 × 5mm; rostellum arms 4mm long.

A S O N D J F M A M J J

Occasional in seeps or on stream banks, from 130–2,000m. Pollinated by beetles or self-pollinating. Hybridizes with *C. grandiflora*. Flowers between November and December, mostly after fire.

Ceratandra grandiflora

Tsitsikamma 13.11.2010

Tsitsikamma 13.11.2010

Tsitsikamma 05.11.2007

Joubertina 15.12.2010

Fairly robust terrestrial up to 430mm tall. **Leaves** numerous, linear to narrowly triangular, up to 80 × 6mm. **Inflorescence** head-like to flat-topped, up to 70mm long, many-flowered; bracts with minute hairs. **Flowers** non-resupinate, sepals pale green, petals and lip orange-yellow flushed with red. Median sepal narrowly spear-shaped, fused to petals, 11–12 × 3–4mm; lateral sepals 10–12 × 5–6mm; petals 12–14 × 7–8mm; lip 6–8 × 8–10mm, with 2 broad, rounded, recurved lobes; rostellum arms 6mm long.

A S O N D J F M A M J J

Occasional in damp habitats, from near sea level to 800m. Pollinated by monkey beetles. Hybridizes with *C. atrata* and *C. globosa*. Flowers between October and December, after fire.

Ceratandra harveyana

Syn: *Evota harveyana*

Betty's Bay 29.11.2008

Betty's Bay 29.11.2008

Betty's Bay 03.12.2008

Betty's Bay 29.11.2008

Slender terrestrial up to 210mm tall. **Leaves** linear to narrowly spear-shaped, up to 40 × 4mm. **Inflorescence** lax to fairly dense, up to 90mm long, 2–12-flowered. **Flowers** with sepals pale green, petals pale yellow, lip yellow. Median sepal 9–11 × 3–4mm; lateral sepals 9–12 × 5–7mm; petals 9–12 × 7–9mm; lip anchor-shaped, 5 × 5mm; lip appendage a pair of fleshy wings, 5 × 5mm; rostellum arms curving around base of lip appendage, 5mm long.

A S O N D J F M A M J J

Rare in fynbos, from 15–900m. Flowers between November and December, after fire.

Ceratandra venosa

Betty's Bay 15.11.2008

Betty's Bay 15.11.2008

Betty's Bay 15.11.2008

Betty's Bay 15.11.2008

Slender terrestrial up to 260mm tall. **Leaves** linear to narrowly spear-shaped, up to 70 × 5mm. **Inflorescence** dense, up to 180mm long, many-flowered. **Flowers** with sepals green suffused with pink, petals and lip white, variously suffused with red-pink, lip with green basal process. Median sepal 4–6 × 2mm; lateral sepals 7 × 2–4mm; petals 6 × 5mm; lip square to triangular or rounded, notched with small, regular teeth, 5 × 7mm; lip appendage with a pair of erect, laterally flattened, rounded horns, 4–5mm tall; rostellum lobes curving around base of horns on lip appendage, 5mm long.

A S O N D J F M A M J J

Very rare on dry Table Mountain Group slopes, from near sea level to 1,000m. Flowers between October and November, after fire.

Cheirostylis nuda

Syn: *Cheirostylis gymnochiloides*

Mtunzini 03.09.2012

Port Durnford 19.08.2012

Mtunzini 28.07.2013

Port Durnford 30.07.2012

Port Durnford 28.07.2012

Terrestrial up to 300mm tall. **Leaves** 4 or 5, along lower part of the erect stem, stalked, 5–7-veined, egg-shaped, 25–65 × 10–30mm. **Inflorescence** dense, glandular-hairy, 20–30 × 15mm, many-flowered. **Flowers** ochre-yellow to white. Sepals and petals similar, glandular-hairy on the outside, 3–4 × 1–1.5mm; lip deeply concave in basal half with fleshy calli and the apex broadly 2-lobed, 2.5–3.5 × 1.5–2mm.

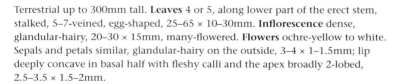

A S O N D J F M A M J J

Locally common in leaf litter in damp to wet, shady forest, from sea level to 300m. Flowers between late July and September.

Corycium alticola

Sehlabathebe (Lesotho) 30.01.2009

Sehlabathebe (Lesotho) 30.01.2009

Naudesnek 03.02.2014

Naudesnek 05.02.2014

Naudesnek 03.02.2014

Robust terrestrial up to 400mm tall. **Leaves** numerous, cauline, spear-shaped, 55–200 × 28mm. **Inflorescence** dense, many-flowered; bracts overtopping flowers. **Flowers** with sepals white turning black, petals pale green, lip pale maroon; 20 × 7–10mm. Median sepal 8 × 4–6mm; lateral sepals deflexed, concave, almost round, fused for half their length, 6–8 × 4–5mm; petals as long as median sepal; lip almost square, notched apically, 5–6 × 4–5mm; lip appendage shield-like, 5mm tall, with 2 strongly deflexed apical lobes.

| A | S | O | N | D | J | F | M | A | M | J | J |

Rare in damp grassland, from 1,950–2,400m. Flowers between January and February.

Corycium bicolorum

Gansbaai 16.11.2009

Hermanus 02.11.2007

Robust to slender terrestrial up to 400mm tall. **Leaves** few to many, narrowly spear-shaped from a broad base, up to 180 × 16mm. **Inflorescence** dense, many-flowered. **Flowers** greenish-yellow, 10 × 4mm. Median sepal 4–5 × 1.5mm; lateral sepals fused, 3–4.5 × 3mm; petals 4 × 3mm; lip shallowly 2-lobed, 1.5–2 × 2mm; lip appendage 2–3mm tall, forming a shield over the column, with 2 small lobes at the sides.

Occasional in sandy areas, from near sea level to 600m. Flowers between October and November, after fire.

Corycium bifidum

Houwhoek 25.10.2010

Gansbaai 11.11.2010

Gansbaai 11.11.2010

Fairly robust terrestrial up to 250mm tall. **Leaves** numerous, cauline, narrowly triangular, sheathing, up to 25 × 8mm. **Inflorescence** dense, many-flowered. **Flowers** with sepals and petals yellowish-green and lip green; 10 × 4mm. Median sepal 4–4.5 × 1.5mm; lateral sepals fused for most of their length, 5mm long; petals 4–6 × 3mm; lip elliptic to triangular, 2.5–4 × 1mm; lip appendage arching forward, broad at apex.

A S O N D J F M A M J J

Very rare in sandy areas in fynbos, from near sea level to 450m. Flowers between October and November, after fire.

Corycium carnosum

Villiersdorp 15.11.2008

Betty's Bay 30.10.2007

Mossel Bay 03.11.2007

Grabouw 14.12.2009

Slender to fairly robust terrestrial up to 550mm tall. **Leaves** numerous, cauline, narrowly spear-shaped, up to 250 × 60mm. **Inflorescence** dense, 4–many-flowered. **Flowers** fleshy, sepals green, petals pink, lip whitish-pink; 18 × 7mm. Median sepal 5–7 × 2–2.5mm; lateral sepals free, 5–7 × 3–3.5mm; petals 6–7 × 3mm; lip 3 × 3.5–4mm, with 2 broad diverging lobes; lip appendage forming a beaked hood over the column, 3–4 × 1.5mm.

A S O N D J F M A M J J

Locally common in fynbos, especially in seepage areas, from near sea level to 1,500m. Pollinated by oil-collecting bees. Flowers between October and January, after fire.

Corycium crispum

Vanrhynsdorp 07.09.2008

Nieuwoudtville 16.09.2007

Vanrhynsdorp 21.09.2009

Vanrhynsdorp 21.09.2009

Nieuwoudtville 21.09.2009

Robust terrestrial up to 400mm tall. **Leaves** numerous, cauline, spear-shaped, margins crisped, up to 140 × 20mm; lower leaves with dark spots at base. **Inflorescence** fairly dense, 5–many-flowered. **Flowers** yellow, lip appendage green; 20 × 5mm. Median sepal 6–7 × 1.5–2mm; lateral sepals fused except for the terminal quarter, 7–8 × 5–6mm; petals with outer third slightly recurved, 7 × 5mm; lip 3mm long, with 2 broad apical lobes; lip appendage 4mm tall, with 2 broad, deflexed apical lobes.

A S O N D J F M A M J J

Occasional in sandy areas, from sea level to 1,500m. Pollinated by oil-collecting bees. Flowers between August and October.

Corycium deflexum

Calvinia 09.10.2009

Calvinia 11.10.2009

Calvinia 01.10.2009

Fairly robust terrestrial up to 250mm tall. **Leaves** numerous, spear-shaped, sometimes withered at flowering time, up to 150 × 14mm. **Inflorescence** lax to fairly dense, 6–20-flowered; bracts with apex withered at flowering time. **Flowers** yellow, lip appendage green, sepals turning brown at flowering time; 12 × 6mm. Median sepal 5–7 × 2–3mm; lateral sepals fused to about halfway, 5–6 × 5mm; petals with outer margin inrolled, 5–7 × 5mm; lip 4–5mm long, broadened and with 2 broad apical lobes 3.5mm wide; lip appendage shield-like, 3–6mm tall, with 2 strongly deflexed, acute apical lobes, each 4 × 2mm.

A S O N D J F M A M J J

Localized and sporadic in short, dry scrub, from 760–1,600m. Pollinated by oil-collecting bees. Flowers between September and October.

Corycium dracomontanum

KZN south coast 19.10.2012

Estcourt 05.02.2007

Sabie 02.01.2011 KZN south coast 19.10.2012

Sentinel 01.03.2008

Slender terrestrial up to 220mm tall. **Leaves** several, cauline, spear-shaped to linear, 50–210 × 15mm, grading into bracts. **Inflorescence** dense, 20–50-flowered. **Flowers** with sepals pale green turning black, petals and lip green tinged with purple, lip appendage bright green; 15 × 7mm. Median sepal 2.5–4 × 2.5–3mm; lateral sepals deflexed, free to fused for up to half their length, 4–5 × 2.5–3mm; petals 4 × 3mm; lip 2–4 × 2.5–4mm; lip appendage shield-like, 5mm tall, clawed with 2 oblong lobes, each 4mm long.

Reliably distinguishable from the very similar *C. nigrescens* only by the lateral processes of the lip appendage, which are nearly oblong and facing sideways.

A S O N D J F M A M J J

Common in grassland, from near sea level to 3,000m. Pollinated by oil-collecting bees. Flowers between October and March.

Corycium excisum

Op-die-Berg 15.11.2009

Op-die-Berg 01.11.2009

Op-die-Berg 12.11.2009

Op-die-Berg 31.10.2009

Slender to fairly robust terrestrial up to 260mm tall. **Leaves** numerous, linear, base sheathing, 120 × 12mm. **Inflorescence** dense, many-flowered. **Flowers** greenish-yellow to yellowish-green, striped, sepals often paler; 8 × 5mm. Median sepal 4–6 × 1.5mm; lateral sepals fused except at apex, 3–5.5 × 4mm; petals 4–5 × 4mm; lip 2.5 × 3mm; lip appendage shield-like with 2 obscure horizontal lobes apically.

A S O N D J F M A M J J

Rare in sandy areas, from near sea level to 800m. Flowers between October and December, after fire.

Corycium flanaganii

Naudesnek 11.12.2008

Naudesnek 11.12.2008

Naudesnek 11.12.2008

Fairly robust terrestrial up to 300mm tall. **Leaves** numerous, spear-shaped, up to 120 × 24mm. **Inflorescence** dense, many-flowered; bracts deflexed, longitudinally striped. **Flowers** non-resupinate, green, lip striped dark maroon; 18 × 8mm. Median sepal 5–5.5 × 3–3.5mm; lateral sepals 7–8 × 3mm; petals similar to lateral sepals; lip 7 × 3mm; lip appendage oblong, sharply notched, 5 × 2mm.

A S O N D J F M A M J J

Uncommon in montane grassland, from 1,300–2,600m. Flowers between November and January.

Corycium ingeanum

Nieuwoudtville 22.09.2009

Nieuwoudtville 16.09.2007

Nieuwoudtville 15.09.2007

Nieuwoudtville 15.09.2007

Fairly robust terrestrial up to 200mm tall. **Leaves** usually 7, spear-shaped, 25–100 × 14–21mm; upper leaves smaller. **Inflorescence** fairly dense, 23–33-flowered. **Flowers** greenish-yellow, sepals turning membranous and brown, petals tipped reddish to black, lip bright green. Median sepal 9 × 4mm; lateral sepals fused, except at apex, 6mm long; petals 10 × 8mm; lip 3mm long, with 2 spreading lobes near apex; lip appendage shield-like, 5–6mm long, with 2 lobes arching backward.

A S O N D J F M A M J J

Localized in sandy or loamy soil in open ground between small karroid shrubs, from 700–800m. Pollinated by oil-collecting bees. Flowers between September and October.

Corycium microglossum

Cape Peninsula 28.11.2008

Cape Peninsula 18.11.2008

Cape Peninsula 28.11.2008

Robust terrestrial up to 365mm tall. **Leaves** numerous, linear from a broad sheathing base, up to 150 × 15mm. **Inflorescence** dense, many-flowered. **Flowers** with sepals and petals greyish-brown and dry, lip appendage green; 24 × 7mm. Median sepal 6–7 × 3mm; lateral sepals fused in lower portion, 5–7 × 6mm; petals 7 × 5mm; lip 7 × 1.5mm; lip appendage broadly elongate to egg-shaped, arching over the column, 5 × 5mm.

A S O N D J F M A M J J

Very rare in sandy soils on mountains and flats, from near sea level to 1,000m. Flowers mainly in November, after fire.

Corycium nigrescens

Witsieshoek 31.01.2007

Verloren Vallei 22.12.2013

Vryheid 28.01.2013

Witsieshoek 01.02.2007

Slender terrestrial up to 550mm tall. **Leaves** several, cauline, narrowly spear-shaped, up to 200 × 24mm, grading into bracts. **Inflorescence** dense, 25–60-flowered. **Flowers** pale green, sepals turning black and then drying with age, petals purple-brown to black, lip maroon to purplish-green; 16 × 6mm. Median sepal 3 × 2–3mm; lateral sepals usually joined for about two thirds of their length, 3.5–5 × 3.5–4mm; petals 3.5–4 × 2.5mm; lip 3 × 3mm; lip appendage shield-like, 3–4mm tall, with 2 deflexed, rolled lobes, each up to 5mm long.

Reliably distinguishable from the very similar *C. dracomontanum* only by the lateral processes of the lip appendage that are tapering and recurved.

A S O N D J F M A M J J

Common in montane grassland, from near sea level to 3,000m. Pollinated by oil-collecting bees. Flowers between December and February.

Corycium orobanchoides

Syn: *Corycium vestitum*

Paarl 26.09.2009

Citrusdal 26.09.2010

Villiersdorp 13.09.2006

Citrusdal 26.09.2010

Slender to robust terrestrial up to 420mm tall. **Leaves** numerous, spear-shaped, up to 180 × 30mm; basal leaves prominently barred with purple-red. **Inflorescence** dense, many-flowered. **Flowers** yellow-green, petal apices purple; 15 × 5mm. Median sepal 5–6 × 1.5–2mm; lateral sepals fused along entire length, 3–5 × 4mm; petals 6–8 × 3–4mm; lip 2–3 × 2.5–3.5mm, broadest apically, with 2 diverging lobes; lip appendage shield-like, 6.5 × 4mm, with lobes horizontal and pointing backward.

A	S	O	N	D	J	F	M	A	M	J	J

Common in sandy areas, often at roadsides or on lawns, from near sea level to 500m. Pollinated by oil-collecting bees. Flowers between September and October.

Corymborkis corymbis

Umtamvuna 12.01.2012

Port Shepstone 02.02.2014

Umtamvuna 01.02.2012

Umtamvuna 09.01.2012

Thin-stemmed, semi-woody, evergreen terrestrial up to 1m tall. **Leaves** spirally arranged, deep glossy green with prominent veins, pleated, elliptic, 110–340 × 30–100mm. **Inflorescence** lax, up to 70mm long, 5–15-flowered. **Flowers** cream to greenish-white, star-shaped. Sepals and petals similar, linear to spatula-shaped, 45–90 × 2–5mm; lip similar, but broadening to 5–13mm near apex.

| A | S | O | N | D | J | F | M | A | M | J | J |

Very local in deep shade in dense forest, from sea level to 700m. Flowers between January and February.

Cynorkis compacta

Kloof 18.08.2012

Kloof 18.08.2012

St Helier 08.09.2013

St Helier 03.09.2013

Very slender lithophyte up to 80mm tall. **Leaves** 1 or rarely 2, basal, sheathing, narrowly egg-shaped, up to 80 × 15mm. **Inflorescence** fairly dense, 1–15-flowered. **Flowers** white, lip spotted pink. Sepals unequal, median broadly elliptic, 3–4.5mm long, laterals spreading, 4–5.5mm long; petals erect, 4mm long; lip 3-lobed, broadly egg-shaped, 6–9 × 6–9mm; lip midlobe square to oblong, sometimes obscurely 3-lobed; lip side lobes 2–3mm long; spur 3–4mm long.

| A | S | O | N | D | J | F | M | A | M | J | J |

Rare in rock crevices on sandstone, from 400–1,500m. Flowers between August and September.

Cynorkis kassneriana

Graskop 06.02.2013

Warburton 04.12.2011

Graskop 06.02.2013

Warburton 18.02.2010

Slender terrestrial, rarely epiphytic, up to 400mm tall. **Leaves** 1 or rarely 2, basal, narrowly elliptic, with 1 to several sheaths above, 80–140 × 25–45mm. **Inflorescence** lax, glandular-hairy, 5–20-flowered. **Flowers** pale blue to mauve, lip sometimes with dark speckles. Sepals unequal, median erect, up to 6mm long, laterals spreading, 7mm long; petals erect, 6mm long; lip 3-lobed, egg-shaped, 9 × 6mm; lip midlobe triangular, 5mm long; lip side lobes 1.5mm long; spur nearly club-shaped, 8–9mm long.

A S O N D J F M A M J J

Common in leaf litter in montane forests and pine plantations, from 1,200–2,000m. Flowers between January and March.

Cyrtorchis arcuata

Oribi Gorge 12.01.2012

Umtamvuna 22.10.2011

Sedgefield 26.11.2013

Umtamvuna 30.09.2012

Robust epiphyte or lithophyte with stems 150–400mm × 5–10mm.
Leaves strap-shaped, folded together basally, unequally bilobed, pale green,
100–160 × 20–30mm. **Inflorescences** 1–3, lateral, usually longer than leaves,
5–15-flowered. **Flowers** white to cream, turning orange with age. Sepals, petals
and lip similar, narrowly spear-shaped, 20–30 × 5–7mm; spur curved, tapering
from a broad entrance to an acute apex, 25–40mm long.

This species is very variable.

Subsp. *arcuata* is widespread in Africa and the only subspecies occurring
in South Africa and Swaziland.

A S O N D J F M A M J J

Common in both montane and lowland forests, from near sea level to 1,500m. Flowers between September
and May.

Cyrtorchis praetermissa

Robust epiphyte with stems 100–300mm × 5–10mm. **Leaves** strap-shaped, dark green, equally or unequally bilobed, in 2 opposite rows, leathery to somewhat succulent, up to 100 × 10mm. **Inflorescences** 1–3, lateral, usually shorter than leaves, 5–15-flowered. **Flowers** white with a greenish tinge, turning orange with age. Sepals, petals and lip similar, 8–13 × 2–4mm; spur tapering, 20–30mm long.

Cyrtorchis praetermissa subsp. *praetermissa*   Syn: *Cyrtorchis praetermissa* var. *praetermissa*

White River 05.01.2014

White River 05.01.2014

Bushbuckridge 15.12.2012

White River 05.01.2014

Subsp. *praetermissa* is distinguished as follows: leaves folded.
　　Subsp. *praetermissa* is widespread in Africa, south of Uganda.

Localized, but often abundant in riverine and montane forest, from 700–1,300m. Flowers between November and January.

A	S	O	N	D	J	F	M	A	M	J	J

Cyrtorchis praetermissa subsp. *zuluensis*

Syn: *Cyrtorchis praetermissa* var. *zuluensis*

KZN north coast 14.01.2012

KZN north coast 28.01.2013

KZN north coast 08.02.2014 KZN north coast 07.02.2014

KZN north coast 08.02.2014

Subsp. *zuluensis* is distinguished as follows: leaves flat.
Subsp. *zuluensis* is restricted to KwaZulu-Natal.

Localized, but often abundant in moist, hot, lowland rainforest and cooler inland forests, from near sea level to 600m. Flowers between January and February.

Diaphananthe fragrantissima

KZN north coast 02.10.2012

KZN north coast 26.02.2014

KZN north coast 26.02.2014

KZN north coast 26.02.2014

KZN north coast 26.02.2014

Robust epiphyte with stems stout, up to 10mm in diameter. **Leaves** thick and leathery, strap-shaped, very unequally bilobed, 150–300 × 20mm. **Inflorescences** usually several, lateral, hanging, 100–300mm long, many-flowered. **Flowers** opposite or in whorls of 4, greenish or yellowish, almost translucent. Sepals spreading forward, 8–12 × 2–3mm; petals arched over the column, 6–10 × 2mm; lip fringed along the tip, 8–13 × 3–6mm, with apiculus 3mm long; spur inflated above the base and laterally compressed, 7–10mm long.

| A | S | O | N | D | J | F | M | A | M | J | J |

Rare and localized in hot coastal and riverine bush, from near sea level to 200m. Flowers between February and March.

Diaphananthe millarii

Syn: *Mystacidium millarii*

Pinetown 21.01.2013

Pinetown 21.12.2013

Pinetown 18.11.2012

Pinetown 21.01.2013

Slender epiphyte; stems up to 40mm long. **Leaves** strap-shaped with conspicuous venation, unequally bilobed, 100–150 × 13–18mm. **Inflorescences** 1 to several, lateral, up to 60mm long, 13–18-flowered. **Flowers** white, anther cap bright green. Sepals spreading forward, 5–7 × 2–3mm; petals spreading, 6 × 4mm; lip entire, opening widely into the tapering 20mm-long spur.

A S O N **D** **J** **F** M A M J J

Rare in coastal forest, from near sea level to 500m. Flowers between December and February.

Didymoplexis verrucosa

KZN north coast 04.09.2013

KZN north coast 04.09.2013

KZN north coast 05.09.2013

KZN north coast 04.09.2013

Slender achlorophyllous terrestrial up to 100mm tall; stem pale brown to white. **Leaves** reduced to a few scattered scales. **Inflorescence** with several terminally aggregated flowers. **Flowers** non-resupinate, opening sequentially, remaining open only for a few hours, white, tinged with brown, lip callus yellow. Sepals equal, rounded, fused near the base with petals, 6–9 × 3–5mm; petals similar, 5–6 × 2–5mm; lip shortly stalked, obscurely 3-lobed, 5–6 × 8–9mm; lip lobes rounded, with callus a strip of papillae along the midline and side lobes pointing upward. Fruiting stalk elongating rapidly after pollination and reaching 100–200mm at maturity.

A S O N D J F M A M J J

Very rare and localized in humus and leaf litter on the floor of dune forest, near sea level. Flowers between August and September.

Disa aconitoides

Weza 19.12.2010   Maclear 14.12.2013   Tsitsikamma 05.11.2007

Tsitsikamma 05.11.2007   Tsitsikamma 05.11.2007

Slender terrestrial 255–600mm tall, rarely producing sterile shoots. **Leaves** on fertile shoot 5–10, cauline, 45–80 × 8–20mm. **Inflorescence** lax, 15–50-flowered. **Flowers** facing downward at 45°, off-white or flushed pale mauve, often with darker mauve spots. Median sepal egg-shaped to elliptic, galea 5–8 × 1.5–3.5mm, 2–3mm deep, the mouth open or the margins bending inward; spur conical, laterally flattened, usually with a rounded apex; lateral sepals spreading, narrowly oblong, 5–7 × 1.5–2.5mm, keeled apically with apiculus shorter than 0.5mm; petals erect, 3.5–5 × 1.5–2.5mm; lip narrowly elliptic, papillate near apex.

A S O N D J F M A M J J

Subsp. *aconitoides* is restricted to South Africa and Swaziland and is the only subspecies occurring in the region.

Occasional in damp grassland, from sea level to 2,200m. Flowers between November and December.

Disa albomagentea

Grabouw 28.11.2009

Grabouw 03.12.2009

Grabouw 03.12.2009

Grabouw 03.12.2009

Slender terrestrial 30–250mm tall. **Leaves** 7–14, cauline, linear, 50–120mm long. **Inflorescence** dense, 6–150-flowered. **Flowers** facing outward, median sepal magenta, lateral sepals, petals and lip white, sometimes tinged with magenta and green. Median sepal hooded, deeply and narrowly grooved on top, 7–9 × 4–5mm; spur grooved, 1–1.5mm long; lateral sepals spreading outward, 5.5–8mm long; petals 4 × 2mm; lip 4.5–7mm long.

A S O N D J F M A M J J

Restricted to the Hottentots Holland Mountains in the Western Cape, where it grows in peaty soils in seeps and marshes, from 1,100–1,350m. Hybridizes with *D. obtusa* subsp. *hottentotica*. Flowers between late November and December, after fire.

Disa alticola

Lydenburg 15.01.2010

Lydenburg 11.01.2013

Lydenburg 01.01.2011

Lydenburg 12.01.2010

Dwarf terrestrial 70–170mm tall. **Leaves** 3 or 4, linear, usually up to 60mm long. **Inflorescence** dense, 10–35-flowered; inflorescence stalk, ovary and bracts beetroot-red. **Flowers** white, galea with red-mauve dot apically, petals with dark red apical blotches, lip lime with basal third white. Median sepal hooded, front margin bending inward, 3–4 × 2.5mm, 1mm deep; spur conical, apex decurved, 1–2mm long, with dorsal groove extending up the galea; lateral sepals projecting away, oblong, 4mm long, with prominent apiculi; petals partially included in galea, almost oblong, 2.5mm long; lip strap-shaped, 3–4mm long.

| A | S | O | N | D | J | F | M | A | M | J | J |

Known only from a few sites where it may be locally common, growing in shallow pans in sand and humus on rock sheets, from 2,000–2,200m. Flowers between December and January.

Disa amoena

Sabie 15.01.2010

Sabie 02.01.2011

Sabie 24.02.2010

Sabie 18.01.2010

Slender to fairly robust terrestrial 300–600mm tall. **Leaves** 7–9, rigid, linear, up to 300mm long. **Inflorescence** lax, 4–14-flowered. **Flowers** light to deep pink, with or without lilac spots, variable even within the same population. Median sepal angled forward, hooded, 15–20 × 6mm; spur horizontal, decurved near apex, cylindrical, 25–35mm long; lateral sepals projecting away, oblong, 13–20mm long, with small, slender apiculi; petals narrowly oblong, 7mm long; lip strap-shaped, projecting away with apex deflexed, 15–20mm long.

A S O N D J F M A M J J

Localized in dry, stony grassland, from 1,800–2,200m. Pollinated by long-proboscid flies. Flowers between January and February.

Disa arida

Herold 15.04.2013

Herold 31.03.2011

Herold 15.04.2013

Herold 31.03.2011

Slender terrestrial 450–550mm tall. **Radical leaves** developing after flowering, linear, somewhat spiralling towards apex, up to 350mm long, often persisting as fibrous sheath; **cauline leaves** sheathing. **Inflorescence** lax, 6–9-flowered. **Flowers** pink or mauve-purple with blue iridescence. Median sepal hooded, apiculate, 8–11mm tall, galea slightly laterally flattened, 3–4mm wide, 4–6mm deep; spur cylindrical, horizontal, decurved near apex, 11–15mm long; lateral sepals projecting away, oblong, 9–11 × 3–6mm, with apiculi 2–3mm long; petals narrowly oblong, 4–5mm tall; lip projecting away, spear-shaped, 6–8mm long.

| A | S | O | N | D | J | F | M | A | M | J | J |

Rare in mature vegetation in the southern parts of the Cape Floral Region, from 450–1,450m. Flowers between March and April.

Disa aristata

Tzaneen 31.01.2006   Tzaneen 26.01.2012

Tzaneen 26.01.2012

Slender, erect or flexuose terrestrial 250–500mm tall. **Leaves** cauline, soft; lower 2 or 3 spreading, narrowly spear-shaped, up to 130 × 10–16mm; upper 2 or 3 mostly sheathing. **Inflorescence** lax, 10–18-flowered. **Flowers** pale to deep pink. Median sepal angled forward, 18–25mm long, galea shallow, 3mm wide, 6mm deep; spur horizontal, cylindrical, 8–18mm long; lateral sepals spreading, strap-shaped, 18–25mm long, with apiculi 3–6mm long; petals spear-shaped, 7–10mm long; lip linear, flexuose, 20–25mm long.

| A | S | O | N | D | J | F | M | A | M | J | J |

Localized on rocky ledges and in rock crevices, from 1,500–2,000m. Flowers between January and February.

Disa atricapilla

Cape Peninsula 11.11.2008

Cape Peninsula 30.11.2008

Cape Peninsula 08.12.2008

Cape Peninsula 16.12.2007

Slender to fairly robust terrestrial 50–300mm tall. **Leaves** cauline, clustered at stem base, spaced out higher up, linear to spear-shaped, 50mm long. **Inflorescence** flat-topped, 3–20-flowered. **Flowers** non-resupinate, median sepal white, lateral sepals white on one side, red on the reverse, and shiny dark maroon to black apically, petals and lip pinkish-maroon suffused with green and speckled. Median sepal reflexed, shallowly hooded, laterally flattened, 10–15 × 3mm, 3mm deep; spur rudimentary; lateral sepals spreading upward, narrowly egg-shaped, apical third usually folded together, 10–17mm long; petals narrowly oblong, 7–10mm long; lip projecting away, linear to narrowly diamond-shaped, with apex curved upward.

Distinguishable from *D. bivalvata* by the shape and colour of the lateral sepals.

A S O N D J F M A M J J

Widespread in seepage areas and damp localities, from near sea level to 2,000m. Pollinated by male thread-waisted wasps. Hybridizes with *D. bivalvata*. Flowers between November and January, after fire.

Disa atrorubens

Op-die-Berg 16.10.2009

Op-die-Berg 16.10.2009

Op-die-Berg 21.10.2009

Op-die-Berg 16.10.2009

Slender terrestrial 100–400mm tall, suffused with beetroot-red. **Cauline leaves** green to beetroot-red, narrowly spear-shaped to spear-shaped, longest near the bottom and grading into the bracts, up to 90mm long. **Inflorescence** lax, 10–18-flowered; bracts green. **Flowers** with sepals and spur beetroot-red, lip and petals almost black. Median sepal narrowly oblong, blunt, shallowly concave, 7–10 × 3–4mm; spur cylindrical, tapering, 15–30mm long; lateral sepals narrowly oblong, reflexed at flowering time, 6–8mm long; petals unevenly egg-shaped, 5–7 × 3–5mm; lip oblong to narrowly oblong, fleshy, 5–7 × 2–4mm.

Distinguishable from *D. ophrydea* by the reflexed lateral sepals and wider lip.

A S O N D J F M A M J J

Rare and restricted to sandy areas, from near sea level to approximately 1,000m. Hybridizes rarely with *D. sabulosa*. Flowers between August and October, after fire.

Disa aurata

Swellendam 21.12.2012

Swellendam 21.12.2012

Swellendam 21.12.2012

Swellendam 21.12.2012

Slender terrestrial 100–600mm tall, spreading by stolons. **Basal leaves** up to 10, spreading, narrowly spear-shaped, up to 140mm long. **Inflorescence** lax, up to 150mm long, 7–25-flowered. **Flowers** golden yellow, median sepal spotted deep red. Median sepal hooded, egg-shaped, 5–12mm long, 3–5mm deep; spur horizontal or slightly hanging, conical, 0.5–2.5mm long; lateral sepals projecting away, broadly egg-shaped, elliptic or oblong, 14–16mm long; petals strap-shaped to narrowly oblong, 4–5mm long; lip projecting away, apex curved upward, linear, 3–4mm long.

A S O N **D J** F M A M J J

Distinguishable from *D. tripetaloides* by the yellow flowers, longer lateral sepals and shorter spur.

Occasional on the banks of perennial streams and burnt, peaty banks along mountain crests, from 500–1,100m. Flowers between December and January, stimulated by fire.

Disa barbata

Syn: *Herschelia barbata*

Malmesbury 10.11.2008

Malmesbury 02.11.2008

Malmesbury 02.11.2008

Malmesbury 10.11.2008

Slender, reed-like terrestrial 250–500mm tall. **Radical leaves** 4–7, linear, mostly dry at flowering time, 150–300 × 1mm; **cauline leaves** 5 or 6, completely sheathing, spear-shaped, 20–40mm long. **Inflorescence** lax, 2–14-flowered. **Flowers** white to very pale blue, veins and lip deeper green to purplish-blue, spur often green. Median sepal angled forward, hooded, egg-shaped, 15–25 × 13–18mm, 8–12mm deep; spur conical, not clearly distinct from galea, straight or rarely upcurved, 1–5mm long; lateral sepals projecting away, narrowly oblong, 15–25mm long; petals with linear limb, 5–6mm long, with apex triangular, incised to divided in two, 3–4 × 3mm; lip soon decurved, egg-shaped, deeply torn, 15mm long.

A S **O N** D J F M A M J J

Very rare on sand flats, from near sea level to 200m. Pollinated by carpenter bees. Flowers between October and November.

Disa basutorum

Tiffendell 27.01.2009

Sentinel 26.01.2008

Sentinel 26.01.2008

Sentinel 26.01.2008

Sentinel 26.01.2008

Slender terrestrial 100–240mm tall. **Leaves** 2–4, stalked, elliptic, 15–45mm long. **Inflorescence** dense, 10–20-flowered. **Flowers** facing downward, colour variable: petals and lip usually yellow or lime-green, sepals and spur lime-green to purple. Median sepal hooded, 4–5 × 3mm, 2.5mm deep; spur cylindrical, straight and ascending slightly from the base, 5–6mm long; lateral sepals spreading, often slightly reflexed, narrowly oblong to tapering, 5mm long; petals narrowly egg-shaped, inside the galea, 4mm long; lip narrowly spear-shaped, 4mm long.

| A | S | O | N | D | J | F | M | A | M | J | J |

Localized, but often abundant on the summit of the Drakensberg, in stony, wet to damp alpine grassland, above 2,800m. Flowers between January and March.

Disa baurii

Syn: *Herschelia baurii*

Dullstroom 16.09.2010

Dullstroom 17.09.2010

Dullstroom 17.09.2010

Dullstroom 22.09.2010

Umtamvuna 23.07.2013

Slender, reed-like terrestrial 200–400mm tall. **Radical leaves** 5–10, produced after flowering, up to 300mm long, narrower than 2mm; **cauline leaves** sheathing, 15–25mm long. **Inflorescence** lax, 2–14-flowered. **Flowers** varying from pale sky-blue to deep purple-blue, lip frequently darker than sepals. Median sepal hooded, egg-shaped, 10–20 × 6–12mm, 5–10mm deep; spur cylindrical, 4–6mm long; lateral sepals projecting away, oblong to rarely spear-shaped, 10–18mm long; petals with linear limb, 8–13mm long and apex variously shaped; lip horizontal, at least at the base, egg-shaped, usually deeply dissected, 10–25mm long.

Very variable in several characters across the entire distribution range.

A S O N D J F M A M J J

Widespread, but never common, in damp to well-drained grassland, from 150–2,000m. Flowers mostly between September and October, and as early as the end of July near the coast.

Disa begleyi

Grabouw 05.01.2010

Grabouw 05.01.2010

Grabouw 05.01.2010

Grabouw 05.01.2010

Small, stout terrestrial 50–100mm tall. **Leaves** cauline, narrowly spear-shaped, all approximately of the same length, 20–35mm long. **Inflorescence** dense, almost flat-topped, 5–15-flowered. **Flowers** non-resupinate, white to pale pink with random maroon spots, underside of sepals suffused with maroon. Median sepal reflexed, concave, elliptic, 6–7mm long; spur rudimentary; lateral sepals nearly curved, oblong, 7–9mm long; petals almost flat, oblong, narrowing towards rear, 6–7mm long; lip projecting away, strap-shaped to narrowly oblong, 3–4mm long.

A S O N D J F M A M J J

Very rare and restricted to mountain peaks on bare, stony slopes, from 1,000–1,500m. Flowers between December and January, after fire.

Disa bifida

Ceres 04.09.2011

Ceres 21.09.2009

Syn: *Schizodium bifidum*

Ceres 02.09.2010

Joubertina 26.08.2011

Ceres 21.09.2009

Slender, straight or nearly flexuose terrestrial 85–300mm tall. **Basal leaves** spreading, egg-shaped, spotted or variegated, 10–20 × 5–10mm; **cauline leaves** spear-shaped. **Inflorescence** 1–7-flowered. **Flowers** pink, lip tooth greenish to purple, lip blade with purple spots on upper surface, petals often darker at the apex. Median sepal shallowly hooded, egg-shaped to oblong, 10–15 × 3–6mm; spur usually upcurved, tapering towards apex, 10–13 × 1–2mm; lateral sepals narrowly oblong, 10–14 × 2.5–5mm; petals strap-shaped, 7–8mm long, with apex usually unequally divided; lip 10–17mm long, with a hanging, thread-like apical tooth 4–8mm long.

A S O N D J F M A M J J

Distinguishable from *D. inflexa* by the larger flowers, broader median sepal, the slender, tapering, normally ascending spur, by the earlier flowering time and its occurrence at lower altitudes.

Fairly common on flat areas of deep sand, from near sea level to 900m. Flowers between late August and October, stimulated by fire.

Disa biflora

Ceres 26.08.2010

Ceres 19.09.2008

Vanrhynsdorp 09.08.2013

Vanrhynsdorp 09.08.2013

Ceres 28.08.2011

Slender, flexuose terrestrial 80–300mm tall. **Basal leaves** spatula-shaped, 6–16 × 3–7mm; **cauline leaves** 1–4, spear-shaped with upper half free and distant from inflorescence stalk. **Inflorescence** 1–6-flowered. **Flowers** white to rarely rose-red, lip pale green. Median sepal linear, recurved, basal part forming a hood over the petals, 7–12 × 1.5–2mm; spur straight or slightly decurved, divided at the tip in two, 5–10mm long; lateral sepals linear to spear-shaped, recurved, 12–18mm long; petals 3–5mm long, with apex divided in two; lip with an apical tooth approximately 8mm long.

Distinguishable from *D. longipetala* by the absence of a long thread-like tooth at the end of the petals.

A S O N D J F M A M J J

Occasional in fynbos and renosterveld, on sandy and clay slopes and flats, from sea level to 900m. Flowers between August and September, stimulated by fire.

Disa bivalvata

Grabouw 02.01.2010

Grabouw 07.12.2009

Houwhoek 09.12.2010

Cape Peninsula 11.11.2008

Slender to fairly robust terrestrial 100–450mm tall. **Leaves** cauline, clustered at the stem base, spaced almost equally higher up the stem, linear to spear-shaped, up to 80mm long. **Inflorescence** dense, flat-topped, 3–30-flowered. **Flowers** non-resupinate, sepals white, petals and lip dark red to almost black. Median sepal reflexed to horizontal, shallowly hooded, laterally flattened, 9–13 × 3–6mm, 2–4mm deep; spur rudimentary; lateral sepals oblong to narrowly egg-shaped, flat to shallowly concave, 10–15mm long; petals oblong, 5–7mm long; lip projecting away, apex often curved upward, strap-shaped, often narrowly diamond-shaped, 5–8mm long.

Distinguishable from *D. atricapilla* by the shape and colour of the lateral sepals.

A S O N D J F M A M J J

Widespread and sometimes locally abundant in slight seepages or on dry mountain slopes, from sea level to 2,000m. Pollinated by spider-hunting wasps. Hybridizes with *D. atricapilla*. Flowers between September and January, after fire.

Disa bodkinii

Grabouw 12.11.2009

Grabouw 12.12.2009

Grabouw 12.11.2009

Grabouw 17.11.2009

Slender to stout terrestrial 30–250mm tall. **Leaves** cauline, spear-shaped, reaching to, or rarely over, the inflorescence. **Inflorescence** flat-topped, 1–15-flowered. **Flowers** non-resupinate, sepals purplish to reddish-brown, petals almost black. Median sepal deeply concave or shallowly hooded, 15 × 9mm, 6mm deep; spur rudimentary; lateral sepals projecting away, spreading forward, narrowly egg-shaped, 15mm long; petals narrowly oblong, 6mm long, with apical calli; lip projecting away, oblong, deeply concave, 9mm long, with a rounded apical callus.

| A | S | O | N | D | J | F | M | A | M | J | J |

Rare in slightly damp areas, from 600–2,000m. Flowers between October and December, after fire.

Disa bolusiana

Grabouw 06.01.2010

Grabouw 06.01.2010

Oudtshoorn 02.12.2007

Misgund 15.01.2011

Oudtshoorn 24.11.2004

Oudtshoorn 02.12.2007

Slender terrestrial 200–300mm tall. **Leaves** narrowly oblong, the larger, lower 2–5 semi-erect, up to 50mm long, grading into sheathing upper leaves. **Inflorescence** lax, 2–25-flowered; bracts visibly net-veined. **Flowers** lime-green, occasionally tinted red. Median sepal erect, shallowly hooded, oblong, 8–11mm long; spur slender, 16–22mm long; lateral sepals reflexed, oblong, 6–8mm long; petals 6–8mm long; lip projecting away or recurved, elliptic, rather fleshy, 5–7mm long.

Distinguishable from *D. comosa* by the leaves that become gradually smaller from the base to the apex of the stem, grading into sheathing upper leaves, and by the later flowering time.

A	S	O	N	D	J	F	M	A	M	J	J

Frequent on high, exposed ridges in gravelly soils, from 200–1,700m. Flowers between November and January, mostly after fire.

Disa brachyceras

Caledon 07.09.2012

Caledon 07.09.2011

Caledon 07.09.2012

Caledon 07.09.2012

Caledon 04.09.2006

Dwarf terrestrial 30–70mm tall. **Leaves** numerous, spiralling, linear, up to 25mm long. **Inflorescence** cylindrical, 5–15-flowered. **Flowers** white, petals and lip tipped with maroon. Median sepal hooded, 2–3mm long; spur not apparent, sac-like; lateral sepals projecting away, oblong, 2.5mm long; petals egg-shaped, 2mm tall; lip strap-shaped, 2mm long.

Distinguishable from *D. tenella* subsp. *pusilla* by the sac-like spur.

A S O N D J F M A M J J

Very rare, in full sun in damp, sandy or occasionally clay soil in renosterveld and fynbos, from 300–600m. Flowers between August and September.

Disa bracteata

Syn: *Monadenia bracteata*

Cape Peninsula 23.09.2007

Cape Peninsula 01.10.2009

Cape Peninsula 26.09.2010

Cape Peninsula 01.10.2009

Houwhoek 09.10.2011

Slender to fairly robust terrestrial 25–300mm tall. **Leaves** numerous, linear to spear-shaped, 40–120mm long. **Inflorescence** cylindrical, 20–120mm long, many-flowered. **Flowers** green, sepals usually tinted maroon. Median sepal shallowly hooded, broadly oblong, 3–4mm long; spur shallowly triangular, hanging, 3–4.5mm long; lateral sepals projecting away, oblong, 2.5–3.5mm long; petals erect, partially included in the galea, 2–2.5mm long; lip hanging, narrowly oblong to strap-shaped, often spear-shaped, 2–2.5mm long.

Distinguishable from *D. densiflora* by the cylindrical spur not constricted at the base and approximately the same length as the median sepal.

A S O N D J F M A M J J

Widespread and common in a variety of habitats, and in areas of mild disturbance, from near sea level to 2,000m. A roadside weed in Australia. Self-pollinating. Flowers between September and November, stimulated by fire.

Disa brevicornis

Syn: *Monadenia brevicornis*

Garden Castle 20.01.2008

Witsieshoek 27.01.2008

KZN south coast 31.08.2013

Witsieshoek 16.01.2008

Garden Castle 23.01.2008

Ramatseliso 03.02.2008

Slender terrestrial 200–500mm tall, with sterile shoots often present. **Leaves** on fertile shoots narrowly spear-shaped, largest near the base, up to 150mm long. **Inflorescence** lax, 20–45-flowered. **Flowers** with petals lime-green, lip lime-green with a maroon base, lateral sepals green, median sepal rust-coloured to maroon. Median sepal shallowly hooded, egg-shaped to oblong, held almost horizontally, 7–10mm tall; spur cylindrical, 2–3 × 7–11mm, with apex curved towards ovary; lateral sepals reflexed, oblong, 5–9mm long; petals egg-shaped to oblong, 5–9mm tall; lip narrowly oblong, 6–8mm long.

A S O N D J F M A M J J

Widespread and locally abundant in montane grassland, although recorded from near sea level to 2,700m. Flowers between late August and February, stimulated by fire.

Disa caffra

Umtamvuna 11.11.2012

Umtamvuna 16.11.2011

Umtamvuna 09.11.2011

Slender terrestrial 250–500mm tall, rarely with a sterile shoot. **Leaves** on fertile shoots cauline, up to 7. **Inflorescence** lax to fairly dense, 8–18-flowered. **Flowers** borne horizontally, plain pink, sometimes reddish or purplish, or purple-spotted, 15–20mm in diameter. Median sepal hooded, 8–12 × 6–10mm, 4–6mm deep; spur slender, cylindrical, gradually decurved, 8–15mm long; lateral sepals projecting away, 8–13 × 5.5–10mm; petals erect, 6–9 × 1.5–3mm; lip narrow, spear-shaped to elliptic, 6–8 × 1.2–2.5mm.

A S O N D J F M A M J J

Rare in swampy areas on the coast, from 100–400m. Flowers between October and November.

Disa cardinalis

Barrydale 14.01.2009

Barrydale 14.02.2012

Barrydale 07.01.2010

Barrydale 07.01.2010

Slender terrestrial 300–600mm tall. **Leaves** 6–10, basal, narrowly elliptic, 50–100mm long. **Inflorescence** lax to fairly dense, 8–25-flowered. **Flowers** pale orange to bright red. Median sepal hooded, 10–15 × 8–10mm, 6–8mm deep; spur conical, 4mm long; lateral sepals projecting away with apex often reflexed, elliptic, 18–28mm long; petals included in the galea, curved over the anther, 6–7 × 1–2mm; lip projecting away, narrowly rhomboid, 6–7 × 1–2mm.

A S O N D J F M A M J J

Localized along streams on the inland slopes of mountains, from 440–750m. Flowers mostly between December and March, but sometimes as early as October after fire.

Disa caulescens

Citrusdal 28.11.2013

Citrusdal 28.11.2010

Citrusdal 07.12.2013

Wellington 24.11.1993

Citrusdal 28.11.2013

Slender terrestrial or lithophyte 100–400mm tall. **Leaves** 4–12, cauline, narrowly spear-shaped, up to 100mm long. **Inflorescence** lax, 5–10-flowered. **Flowers** white, petals striped with maroon. Median sepal shallowly hooded, 6–8mm long, 2mm deep; spur somewhat hanging, often curved slightly downward, cylindrical or conical, 2–3mm long; lateral sepals projecting away, apices usually reflexed, egg-shaped to elliptic, rounded, 7.5–11mm long; petals egg-shaped, notched, 3.5–5mm long; lip projecting away, linear, 4–5mm long.

Distinguishable from *D. tripetaloides* by the cauline leaves and egg-shaped petals striped with maroon.

A S O N D J F M A M J J

Occasional on mossy stream banks or in rock fissures in the mountains, often in half-shade, from 300–1,200m. Pollinated by flies. Hybridizes rarely with *D. tripetaloides* and very rarely with *D. uniflora*. Flowers between November and January.

Disa cephalotes

Slender terrestrial 150–400mm tall. **Leaves** cauline, narrowly spear-shaped to linear; lower leaves often clustered basally, up to 200mm long; upper leaves completely sheathing. **Inflorescence** dense, spherical, 15–many-flowered. **Flowers** white to deep pink, sepals with purple spots near the apices. Median sepal deeply hooded, constricted near the middle, margins bending inward, 5–8 × 3–4mm, 4–5mm deep; spur straight or decurved, 3–6mm long; lateral sepals projecting away, broadly oblong, 5–7mm long, with well-developed apiculi; petals oblong, 4mm long; lip strap-shaped, often slightly widened apically, 4–5mm long.

Disa cephalotes subsp. *cephalotes*

Sentinel 15.01.2008

Sentinel 23.01.2008

Sentinel 23.01.2008

Sentinel 22.01.2014

Subsp. *cephalotes* is distinguished as follows: plants 200–400mm tall; leaves hard; flowers white; found below 2,400m.

Subsp. *cephalotes* is endemic to South Africa and Lesotho.

Distinguishable from *Brownleea galpinii* subsp. *major* by the presence of a conspicuous lip.

Occasional in small populations in dry to damp grassland, or rarely in moss on rocky outcrops in full sun, from 1,500–2,400m. Pollinated by long-proboscid flies. A mimic of *Scabiosa* flowers. Flowers between January and February.

| A | S | O | N | D | J | F | M | A | M | J | J |

Disa cephalotes subsp. *frigida*

Sentinel 26.01.2008

Sentinel 26.01.2008

Sentinel 26.01.2008

Sentinel 22.01.2014

Subsp. *frigida* is distinguished as follows: plants 150–250mm tall; leaves soft; flowers pink; found above 2,400m.

   Subsp. *frigida* is endemic to South Africa and Lesotho.

Rare in dry to damp alpine grassland, from 2,400–3,000m. Flowers between January and March.

A S O N D J F M A M J J

Disa cernua

Syn: *Monadenia cernua*

Gansbaai 22.09.2010

Tsitsikamma 10.10.2008

Tsitsikamma 10.10.2008

Tsitsikamma 10.10.2008

Tsitsikamma 16.10.2012

Robust terrestrial 200–750mm tall. **Leaves** 140–200 × 15–20mm.
Inflorescence lax, 10–40-flowered. **Flowers** with sepals cream-green, mottled
maroon, petals and lip lime-green. Median sepal shallowly hooded, oblong,
10–14mm long; spur club-shaped, 11–17 × 3–5mm, pressed close to the ovary;
lateral sepals reflexed, narrowly oblong to oblong, 9–13mm long; petals more
or less oblong, 7–10mm long; lip strap-shaped, almost fleshy, 8–12mm long.

Distinguishable from *D. physodes* by a spur longer than 10mm.

| A | S | O | N | D | J | F | M | A | M | J | J |

Very rare in marshy fynbos on the flats between the Cape Fold Mountains and the sea, from near sea level to 300m.
Flowers between September and October, after fire.

Disa chrysostachya

Tsitsikamma 05.11.2007

Tsitsikamma 14.11.2011

Tsitsikamma 13.11.2010

Tsitsikamma 14.11.2011

Tsitsikamma 20.10.2014

Robust terrestrial 250mm–1.1m tall, rarely producing a sterile shoot. **Leaves** on fertile shoots 10–20, tightly overlapping, up to 130mm long. **Inflorescence** dense, cylindrical, 40–120-flowered. **Flowers** orange, rarely yellow, 8mm in diameter. Sepals 6–11mm long, median erect; spur hanging, club-shaped, flattened, reaching below base of median sepal, 5–11mm long; petals curved over anther; lip 5–7mm long.

Distinguishable from *D. polygonoides* and *D. woodii* by the long, club-shaped spur reaching below the base of the median sepal.

A S O N D J F M A M J J

Locally common in damp or marshy areas, from near sea level to 2,200m. Hybridizes rarely with *D. rhodantha*. Pollinated by sunbirds. Flowers between November and January.

Disa clavicornis

Sabie 25.01.2014

Sabie 25.01.2014

Moderately robust terrestrial 280–400mm tall. **Leaves** cauline, up to 100mm long. **Inflorescence** dense, 20–50-flowered. **Flowers** facing downward, salmon to pale pink. Median sepal hooded, 5.5mm tall, 4mm deep; spur ascending to erect, slightly curved, slender, cylindrical, 7mm long, with apex strongly club-shaped; lateral sepals spreading, narrowly oblong, 6 × 2.8mm; petals erect, narrowly strap-shaped, 6 × 1mm; lip narrowly strap-shaped, 4.5 × 0.5mm.

| A | S | O | N | D | J | F | M | A | M | J | J |

Very rare in marshes and damp grassland, from 1,900–2,250m. Flowers between January and February.

Disa cochlearis

Ladismith 04.02.1997

Ladismith 04.02.1997

Ladismith 04.02.1997

Slender terrestrial up to 450mm tall, with wiry stem. **Radical leaves** developing after flowering, linear, folded towards apex, up to 50mm long; **cauline leaves** forming papery sheaths. **Inflorescence** lax, 3–11-flowered. **Flowers** white, sometimes tinged with pale mauve. Median sepal with a narrowly elliptical entrance, 11mm tall; spur somewhat ascending from a conical base, tapering, 19mm long; lateral sepals spear-shaped, 10mm long, with apiculi 2mm long; petals 7mm tall; lip linear, 8mm long, with a maroon, warty tubercle near apex.

| A | S | O | N | D | J | F | M | A | M | J | J |

Rare and known only from a single population occurring in renosterveld on the arid Elandsberg, at 1,450m. Flowers between January and February.

146 *Disa*

Disa comosa

Syn: *Monadenia comosa*

Citrusdal 20.09.2009

Franschhoek 04.09.2010

Franschhoek 18.09.2009

Franschhoek 03.09.2010

Slender, erect or nearly flexuose lithophyte 80–300mm tall. **Basal leaves** 2 or 3, elliptic, 30–120mm long; **cauline leaves** closely sheathing the stem. **Inflorescence** lax, 1–20 flowered; bracts visibly net-veined. **Flowers** lime-green, occasionally tinted red. Median sepal shallowly hooded, oblong, 9–11mm tall; spur slender, cylindrical, pressed close to the ovary, 17–24mm long; lateral sepals oblong to egg-shaped, 6–7mm long; petals 6–8mm long; lip elliptic, almost fleshy, 6–8mm long.

Distinguishable from *D. bolusiana* by the two large basal leaves sharply differentiated from the remaining ones, and by the earlier flowering time.

| A | S | O | N | D | J | F | M | A | M | J | J |

Widespread in rock crevices and on ledges, from 500–1,600m but mostly found below 1,000m. Flowers between September and November, with most records from October, stimulated by fire.

Disa conferta

Cape Peninsula 05.10.2007

Houwhoek 28.10.2010

Cape Peninsula 05.10.2007

Cape Peninsula 18.10.2013

Cape Peninsula 18.10.2013

Cape Peninsula 18.10.2013

Slender terrestrial 80–220mm tall, usually suffused with beetroot-red. **Leaves** numerous, linear, up to 70mm long. **Inflorescence** dense, cylindrical, 30–130 × 10mm, many-flowered. **Flowers** lime-green to beetroot-red. Median sepal shallowly hooded, oblong, 2.5mm long; spur pouch-shaped, 0.1–0.2mm long; lateral sepals reflexed, oblong to narrowly egg-shaped, 2.5mm long; petals 2mm long; lip hanging, 2–2.5mm long.

A S O N D J F M A M J J

Occasional on slightly moist sandy or gravelly flats, from sea level to 600m. Flowers between September and October, after fire.

Disa cooperi

Witsieshoek 25.01.2008

Witsieshoek 25.01.2008

Witsieshoek 16.01.2008

Robust terrestrial 400–700mm tall, with sterile shoot 50–100mm tall. **Leaves** on sterile shoot 2–5, up to 250 × 50mm; on fertile shoot 6–15, cauline, up to 200 × 40mm. **Inflorescence** dense, 20–60-flowered. **Flowers** facing downward at 45°, pale to dark pink, occasionally white or cream, lip lime-green; 15mm in diameter. Median sepal hooded, 9–12 × 5–7mm, 4–6mm deep; spur ascending, slender, straight or slightly decurved, 35–50mm long; lateral sepals completely reflexed at flowering time, oblong, 10–14 × 6–8mm, with apiculi up to 2mm long; petals mostly included in galea, 7–10 × 3.5–4.5mm; lip hanging, broadly spatula-shaped, 9–13mm long, with a short claw.

A S O N D J F M A M J J

Locally common in dry to damp high-altitude grassland, from 1,450–2,200m. Pollinated by hawkmoths. Flowers between December and February.

Disa cornuta

Cape Peninsula 27.10.2007

Plettenberg Bay 05.11.2007

Kareedouw 14.09.2013

Kareedouw 14.09.2013

Robust terrestrial up to 1m tall. **Leaves** cauline, spear-shaped to narrowly egg-shaped, usually barred with red basally. **Inflorescence** 15–35-flowered. **Flowers** slightly downward-facing, colour variable, typically white with a purple galea and purple blotch on the lip, but also mauve-lilac or silvery green. Median sepal apiculate, deeply hooded, 12–18mm tall, 8–12mm deep; spur cylindrical, gently or rarely sharply decurved, shallowly grooved below and sharply notched, 10–20mm long; lateral sepals projecting away, narrowly to broadly oblong, 12–16mm long; petals strap-shaped; lip projecting away, 5–10mm long. Very variable in colour and shape of petal apices, lateral sepals and lip over its entire range.

A S O N D J F M A M J J

Widespread in the winter- and summer-rainfall regions, and locally common in well-drained areas in full sun in grassland, and on mountains and sandy flats, from near sea level to 2,200m. Flowers between September and February.

Mamre 19.10.2009

Witsieshoek 14.01.2008

Naudesnek Pass 24.01.2010

Witsieshoek 15.01.2008

Disa crassicornis

Elliot 23.02.2009

Howick 08.12.2011

Witsieshoek 25.01.2008

Robust terrestrial up to 1m tall, rarely producing a sterile shoot. **Leaves** on fertile shoots cauline, largely reduced and sheathing. **Inflorescence** semi-lax, 5–25-flowered. **Flowers** white to cream with pink to purple mottling, up to 50mm in diameter. Median sepal hooded, 20–40 × 15–30mm, 10–15mm deep; spur grading into the galea, cylindrical, decurved, 30–40mm long; lateral sepals spreading horizontally, spear- to egg-shaped, 25–30 × 10–20mm; petals erect, often included in the galea, 20–30 × 10–15mm; lip horizontal basally, soon decurved, narrowly to broadly elliptic, 20–25 × 4–15mm. Very variable in flower shape with high-altitude plants having narrow, tapering flower parts and low-altitude plants having wide, blunt flower parts.

Distinguishable from *D. thodei* by the lateral sepals longer than 20mm.

A	S	O	N	D	J	F	M	A	M	J	J

Occasional in grassland, usually in damp areas, rarely on rock ledges, from 1,000 –2,700m. Pollinated by hawkmoths. Flowers between November and March.

Disa cylindrica

Cape Peninsula 23.09.2007

Grabouw 28.11.2009

George 02.12.2007

Grabouw 16.10.2008

Slender terrestrial 80–350mm tall. **Leaves** cauline, strap-shaped, 30–80mm long. **Inflorescence** cylindrical, many-flowered. **Flowers** green, petals and lip lime-green. Median sepal oblong, concave, 4–6mm long; spur barely visible, 0.2–1mm long; lateral sepals oblong, reflexed, 4–5mm long; petals narrowly egg-shaped to egg-shaped, 3–4mm long; lip horizontal basally, soon deflexed, narrowly elliptic, fleshy, 3.5–4.5mm long.

A S O N D J F M A M J J

Widespread and rather common in damp or swampy areas in fynbos, from sea level to 1,200m. Flowers between late September and December, after fire.

Disa densiflora

Syn: *Monadenia densiflora*

Tsitsikamma 10.10.2008

Cape Peninsula 05.10.2007

Tsitsikamma 05.10.2010

Tsitsikamma 14.10.2013

Tsitsikamma 14.10.2013

Slender terrestrial 75–195mm tall. **Leaves** numerous, linear to spear-shaped, 50–140mm long. **Inflorescence** slender, cylindrical, 29–180mm long, many-flowered. **Flowers** with petals and lip dull green, sepals rusty red to green. Median sepal shallowly hooded, oblong, 3–5mm long; spur triangular in cross-section, constricted at the base, 2–3mm long; lateral sepals oblong to egg-shaped, projecting away, 3–5mm long; petals narrowly oblong, nearly curved, 2.5–5mm long; lip hanging, narrowly oblong to spear-shaped, 2.5–5mm long.

Distinguishable from *D. bracteata* by the spur, which is constricted at the base, triangular in cross-section and approximately half the length of the median sepal.

| A | S | O | N | D | J | F | M | A | M | J | J |

Occasional in sandy soils in a wide range of habitats, from near sea level to 1,000m. Flowers between October and November, after fire.

Disa dracomontana

Sentinel 02.01.2009

Sentinel 02.01.2009

Witsieshoek 01.01.2009

Witsieshoek 01.01.2009

Slender terrestrial 150–450mm tall. **Leaves** cauline, rigid, up to 180mm long.
Inflorescence lax, 5–15-flowered. **Flowers** white to lilac, with conspicuous
darker venation pattern, petals white with purple apices. Median sepal shallowly
hooded, 7–12mm tall; spur horizontal, usually straight, 2.5–5mm long; lateral
sepals oblong to broadly elliptic, 8–13mm long, with long, pointed apiculi;
petals erect inside the galea, spear-shaped, 3.5–5mm long; lip projecting away,
strap-shaped to oblong, 7–9mm long.

   Distinguishable from *D. stricta* by the larger flowers with a clear venation
pattern and longer lateral sepals.

A	S	O	N	D	J	F	M	A	M	J	J

Very rare on grassy slopes, often among rocks or in rocky soils, from 2,100–2,700m. Flowers between December
and February.

Disa draconis

Mamre 10.11.2008

Mamre 01.11.2009

Mamre 14.11.2008

Mamre 16.11.2008

Slender terrestrial up to 600mm tall. **Leaves** cauline; lower ones usually clustered basally, strap-shaped to linear, withered at flowering time, up to 200mm long; upper ones dry, completely sheathing. **Inflorescence** lax, 10–20-flowered. **Flowers** cream or white, galea and petals with purple markings. Median sepal narrowly oblong, basal part hooded, apex reflexed, 15–25mm long; spur cylindrical, 35–45mm long; lateral sepals narrowly oblong, apex reflexed, 15–22mm long; petals strap-shaped, included in the median sepal galea; lip very narrowly spear-shaped, apex deflexed, 15–20mm long.

Distinguishable from *D. karooica* by the petal apices, which are not lobed. Distinguishable from *D. harveyana* by the petal apices, which are included in the galea.

Rare on sandy flats near the coast, from near sea level to 150m. Pollinated by long-proboscid flies. Flowers between October and November.

Disa elegans

Porterville 26.11.2009

Ceres 07.11.2007

Porterville 06.12.2010

Porterville 26.11.2009

Fairly robust terrestrial 200–600mm tall. **Leaves** cauline, linear to spear-shaped, up to 250mm long. **Inflorescence** flat-topped, up to 30-flowered; bracts reaching the flowers. **Flowers** white, petals and lip with apical third maroon and apex rich yellow. Median sepal shallowly hooded, 13–18mm tall, 10mm wide, 5mm deep; spur rudimentary; lateral sepals broadly oblong, concave, 15–22mm long; petals 6–8mm long, with apex curved up behind the anther; lip projecting away, concave, oblong to broadly elliptic, 10mm long.

A S O N D J F M A M J J

Rare, but often in substantial populations in black, peaty soil in marshes, from 1,000–2,000m. Pollinated by scarab beetles. Flowers between November and December, after fire.

Disa esterhuyseniae

Citrusdal 18.04.2010

Paarl 17.02.2009

Citrusdal 18.04.2010

Slender, grass-like terrestrial up to 420mm tall. **Radical leaves** up to 10, developing after flowering, blades strap- to spear-shaped, 20–30mm long, margins wavy, leaf stalks 30–50mm long; **cauline leaves** completely sheathing, 20mm long. **Inflorescence** lax, 10–20-flowered. **Flowers** yellow-green suffused with brown. Median sepal hooded, 4mm tall, 2mm deep; spur cylindrical, horizontal, 6mm long, with apex club-shaped and usually curved upward; lateral sepals spear-shaped, shortly apiculate, 6mm long; petals oblong, 2mm tall; lip linear, 3mm long.

A S O N D J F M A M J J

Rare in slightly damp, stony soil, from 900–1,500m. Flowers between January and April.

Disa extinctoria

Graskop 22.12.2010

Graskop 22.12.2010

Graskop 22.12.2010

Graskop 08.01.2011

Slender terrestrial 300–600mm tall, producing a sterile shoot. **Leaves** on sterile shoots 2 or 3, semi-erect; on fertile shoots 8–12, cauline. **Inflorescence** dense, 30–70-flowered. **Flowers** facing downward, scarlet, petals and lip white or yellow. Median sepal hooded, 4.5–5.5 × 5mm, 4mm deep; spur soon decurved, 3–4.5mm long; lateral sepals projecting away or descending parallel to ovary, oblong, 4.5–5.5 × 2.5mm, with apiculi well-developed; petals erect, 3.5–4.5 × 2.5mm; lip linear, 4.5–5.5mm long.

A S O N D J F M A M J J

Rare in damp grassland and swamps, from 1,000–1,600m. Flowers between December and January.

Disa fasciata

Houwhoek 08.10.2010

Grabouw 05.10.2010

Grabouw 16.10.2010

Grabouw 17.10.2010

Slender terrestrial 50–250mm tall. **Leaves** cauline, mostly sheathing, egg-shaped, 10–20mm long, with sheath usually barred with purple. **Inflorescence** flat-topped, 1–4-flowered. **Flowers** non-resupinate, white with some purplish spots, almost radially symmetrical with sepals and lip borne in the same plane. Median sepal rounded, 9–14mm long, with lower third deeply hooded and apical two thirds flat and reflexed; spur parallel to ovary, slender, 3–5mm long; lateral sepals projecting away, flat, broadly oblong, 8–13mm long; petals almost square in outline, very unequally 2-lobed, middle of inner surface with numerous small papillae, 5mm tall; lip projecting away, oblong to broadly egg-shaped, 8–10mm long.

A S O N D J F M A M J J

Occasional in sandy areas, from near sea level to 1,600m. Flowers in October, after fire.

Disa ferruginea

Cape Peninsula 10.03.2014

Cape Peninsula 06.03.2008

Mossel Bay 25.03.2011

Cape Peninsula 16.03 2013

Stout, reed-like terrestrial 200–450mm tall. **Radical leaves** linear, developing after flowering; **cauline leaves** dry, sheathing. **Inflorescence** dense, 1–40-flowered. **Flowers** bright red to orange, often with some parts yellow. Median sepal apiculate, galea 8–10mm tall, 8–10mm deep; spur slender, grading into the galea, 7–20mm long; lateral sepals projecting away, elliptic to narrowly elliptic, with apiculi up to 4mm long; petals spear-shaped, 5–7mm long; lip narrowly egg- to spear-shaped, 10–12mm long.

| A | S | O | N | D | J | F | M | A | M | J | J |

Occasional to common in dry to slightly damp localities, usually in the zone of the southeaster clouds, from 400–1,500m. Pollinated by the mountain pride butterfly. Hybridizes very rarely with *D. graminifolia*. Flowers between February and March, stimulated by fire.

Disa filicornis

Ceres 01.11.2007

Ceres 01.11.2007

Ceres 16.11.2008

Joubertina 14.11.2010

Slender to fairly robust terrestrial 50–250mm tall. **Radical leaves** numerous, linear, 10–40mm long; **cauline leaves** spear-shaped, almost completely sheathing. **Inflorescence** lax, 1–10-flowered. **Flowers** varying from almost white to bright pink or purplish, galea often darker and spotted or veined on the inside. Median sepal spatula-shaped, with claw 4–6mm long and blade broadly egg-shaped, concave, reflexed, 6–12 × 6–12mm; spur rudimentary; lateral sepals projecting away, oblong, apex often reflexed, 10–20mm long, with apiculi well-developed; petals strap-shaped to oblong, 7–12mm long; lip hanging, linear, 6–12mm long.

Widespread and fairly common on dry mountain slopes, usually in full sun, from 250–1,500m. Pollinated by leafcutter bees. Hybridizes rarely with *D. tenuifolia*. Flowers between October and December, after fire.

Disa flexuosa

Syn: *Schizodium flexuosum*

Op-die-Berg 19.09.2009

Op-die-Berg 19.09.2009

Op-die-Berg 01.09.2010

Op-die-Berg 19.09.2009

Slender, flexuose terrestrial 150–350mm tall. **Basal leaves** stalked, spear- to spatula-shaped, 10–20 × 5–9mm; **cauline leaves** 2–6, spear-shaped, 10–20mm long. **Inflorescence** 1–6-flowered. **Flowers** with sepals white, petals and lip yellow, lip with black spots. Median sepal egg-shaped, 7–11 × 4–9mm; lateral sepals egg-shaped, 7–11 × 4–9mm; petals 3–5mm long with apex divided in two, front tooth as long as the limb, rear tooth very reduced or absent; lip with wavy margins and a wedge-shaped apical tooth.

| A | S | O | N | D | J | F | M | A | M | J | J |

Increasingly rare in seasonally moist, sandy habitats, generally below 1,000m. Flowers between September and October, stimulated by fire.

Disa fragrans

Naudesnek 16.02.2008

Lydenburg 13.03.2013

Sentinel 01.02.2007

Naudesnek 16.02.2008

Naudesnek 06.02.2008

Lydenburg 23.03.2010

Robust terrestrial 70–500mm tall, producing a sterile shoot. **Leaves** intensely to almost invisibly mottled with darker spots; on sterile shoots usually 2, narrowly elliptic; on fertile shoots 4–9, cauline, 30–100 × 15–45mm. **Inflorescence** dense, 30–150-flowered. **Flowers** white to pale lilac or yellow to olive, usually with dark mottling, strongly scented, 8mm in diameter. Sepals 4.5–7mm long; spur hanging, reaching below base of median sepal, slender or apically club-shaped, 4–10mm long; petals curved over the anther; lip 3–7 × 0.4–1mm.

Subsp. *fragrans* is widespread from Tanzania southwards and is the only subspecies occurring in South Africa and Lesotho.

A	S	O	N	D	J	F	M	A	M	J	J

Occasional in high-altitude, rocky grassland in full sun, from 1,800–3,000m. Pollinated by bees, flies and beetles. Hybridizes rarely with *D. sankeyi*. Flowers between January and April.

Disa galpinii

Maclear 24.02.2008

Maclear 14.03.2010

Maclear 21.02.2009

Slender terrestrial 300–600mm tall. **Leaves** cauline, erect, narrowly spear-shaped, up to 120mm long. **Inflorescence** dense, many-flowered. **Flowers** with lateral sepals pale lilac, median sepal white, lip lime-green. Median sepal shallowly hooded, 4–5mm tall, 2mm deep, with 2 sacs at the rear; spur slender, pointing upward, 12–16mm long; lateral sepals reflexed, oblong to egg-shaped, 4–6mm long; petals included in galea, 3mm long; lip hanging, narrowly oblong, 5mm long.

A S O N D J F M A M J J

Very rare in grassland on the Drakensberg, at approximately 2,000m. Flowers between February and March.

Disa gladioliflora

Slender terrestrial 250–500mm tall. **Radical leaves** developing after flowering, linear, up to 300mm long, often persisting as a fibrous sheath; **cauline leaves** sheathing. **Inflorescence** lax, 5–12-flowered. **Flowers** white to pink, often with darker markings. Median sepal hooded, 8–20mm tall, galea laterally flattened, angled slightly forward, 2mm wide, 3–5mm deep; spur cylindrical, horizontal, decurved near the tip, 6–14mm long; lateral sepals projecting away, narrowly oblong to oblong, 8–16mm long; petals 2.5–5mm long; lip projecting away, spear- to egg-shaped, 8–16mm long.

Disa gladioliflora subsp. capricornis

Barrydale 02.02.2014

Barrydale 14.02.2012

Barrydale 15.02.2012

Barrydale 14.02.2012

Subsp. *capricornis* is distinguished as follows: median sepal shorter than 11mm; spur 6–11mm long.

Subsp. *capricornis* is endemic to the Langeberg between Swellendam and Riversdale.

Locally frequent in dry to peaty soil in full sun, from 750–1,300m. Flowers between February and March, stimulated by fire.

A S O N D J F M A M J J

Disa gladioliflora subsp. *gladioliflora*

Mossel Bay 25.03.2011

Mossel Bay 26.03.2011

Mossel Bay 26.03.2011

Mossel Bay 25.03.2011

Subsp. *gladioliflora* is distinguished as follows: median sepal at least 11mm long; spur at least 8mm long.

Subsp. *gladioliflora* is endemic to the western Cape Floral Region, east of Riversdale.

Occasional, singly or in small populations in damp, peaty soil on south-facing slopes, from 800–1,300m. Flowers between March and May, but sometimes as early as the end of December.

| A | S | O | N | D | J | F | M | A | M | J | J |

Disa glandulosa

Grabouw 20.12.2009

Grabouw 20.12.2009

Cape Peninsula 19.12.2007

Grabouw 05.01.2010

Slender terrestrial 60–200mm tall, covered in glandular hairs. **Basal leaves** 3–6, spreading, elliptic, up to 40mm long; **cauline leaves** almost entirely sheathing. **Inflorescence** flat-topped, 1–5-flowered. **Flowers** pink, sepal bases and petals with red spots. Median sepal blunt, shallowly hooded, 5–6mm tall; spur nearly conical, parallel to ovary, 2–3mm long; lateral sepals projecting away, 5–6mm long; petals narrowly oblong, 3–4mm long; lip oblong to spear-shaped, 4mm long.

Distinguishable from *D. vaginata* by its covering of glandular hairs.

A S O N D J F M A M J J

Occasional, and usually in small clusters, in wet moss or turf on rocks and in crevices, from 400–1,500m. Self-pollinating. Flowers between December and January, after fire, but annually in protected rock crevices.

Disa graminifolia

Syn: *Herschelia graminifolia*

Cape Peninsula 05.03.2008

Cape Peninsula 10.03.2014

Cape Peninsula 04.03.2008

Cape Peninsula 04.03.2008

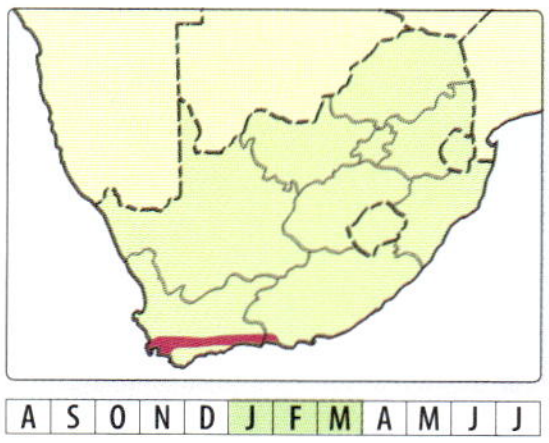
A S O N D J F M A M J J

Slender, reed-like terrestrial 500mm–1m tall. **Radical leaves** usually 5, semi-erect, frequently rolled, 200–500 × 5mm; **cauline leaves** 6–9, closely sheathing, 20–40mm long. **Inflorescence** lax, 2–6-flowered. **Flowers** blue to violet or purple, petal apices green, lip dark purple. Median sepal erect, hooded, 15–20mm tall, 5–10mm deep; spur usually straight, club-shaped, 2–4mm long; lateral sepals projecting away, narrowly oblong to oblong, 13–18 × 6–10mm; petals with strap-shaped limb 11–16mm long and apices expanded into fan-shaped structures, 4–6mm in diameter; lip projecting away, narrowly elliptic to elliptic, margins usually finely toothed and down-curved, 11–16mm long.

Distinguishable from *D. purpurascens* by the nearly club-shaped spur and down-curved lip margins.

Occasional on dry mountain slopes, usually on rocky soils, often in dense vegetation, from 300–1,200m. Pollinated by carpenter bees. Hybridizes very rarely with *D. ferruginea*. Flowers between January and March.

Disa hallackii

Gansbaai 24.10.2010

Gansbaai 24.10.2010

Plettenberg Bay 05.11.2007

Gansbaai 24.10.2010

Robust terrestrial up to 500mm tall. **Leaves** cauline, narrowly spear-shaped to spear-shaped, barred and spotted with red basally. **Inflorescence** dense, 15–35-flowered; bracts prominent, overtopping and partially obscuring the flowers. **Flowers** horizontal, sepals mostly green suffused with purple and white on the inside, dorsal sepal with dark veins, petals purplish, lip white with a purple stripe or blotch. Median sepal hooded, 15–18mm tall, galea 4mm deep; spur almost hanging, sharply decurved, notched and slightly club-shaped, 5mm long; lateral sepals 10–16mm long; petals strap-shaped, 6mm long; lip spear-shaped to narrowly oblong, 8–11mm long.

A S O N D J F M A M J J

Rare along the coast, usually in recent sand, from near sea level to 80m. Flowers between October and November, stimulated by fire.

Disa harveyana

Slender terrestrial up to 600mm tall. **Basal leaves** strap-shaped, dry at flowering, up to 200mm long; **cauline leaves** dry, sheathing. **Inflorescence** lax, 5–10-flowered. **Flowers** cream or mauve, petals with purple or red streaks, galea with purple dots. Median sepal flat, hooded at the base, 15–30mm long; spur slender, 20–90mm long; lateral sepals narrowly strap- to spear-shaped, 15–25mm long; petals strap-shaped, 15–30mm long; lip strap-shaped.

Disa harveyana subsp. harveyana

Cape Peninsula 11.01.2009

Wellington 27.11.2010

Cape Peninsula 20.12.2007

Cape Peninsula 10.12.2010

Subsp. *harveyana* is distinguished as follows: spur shorter than 50mm; later flowering time.

Subsp. *harveyana* is endemic to the Western Cape of South Africa.

Distinguishable from *D. draconis* and *D. karooica* by the petal apices, which are exserted from the galea.

Occasional in shrubby habitats, on relatively dry slopes associated with sandstones, from 300–1,500m. Pollinated by long-proboscid flies. Flowers between late November and January.

| A | S | O | N | D | J | F | M | A | M | J | J |

Disa harveyana subsp. *longicalcarata*

Ceres 07.11.2007

Ceres 28.11.2010

Ceres 02.12.2008

Ceres 07.11.2007

Subsp. *longicalcarata* is distinguished as follows: spur longer than 40mm; earlier flowering time.

Subsp. *longicalcarata* is endemic to the Western and Northern Cape of South Africa.

Localized, but often common on the hot, dry lower slopes of sandstone mountains, often in half-shade under shrubs and sometimes naturalized under alien pines, from 400–1,000m. Pollinated by long-proboscid flies. Flowers between October and November.

| A | S | O | N | D | J | F | M | A | M | J | J |

Disa hians

Syn: *Herschelia hians*

Plettenberg Bay 07.12.2008

Tsitsikamma 17.12.2010

Tsitsikamma 16.12.2011

Tsitsikamma 17.12.2010

Plettenberg Bay 27.11.2012

Slender, reed-like terrestrial 400–600mm tall. **Radical leaves** 8–13, often developing after flowering, semi-rigid, up to 2mm wide; **cauline leaves** completely sheathing, 20–40mm long. **Inflorescence** lax, 3–16-flowered. **Flowers** off-white or very pale blue to purplish-blue, lip frequently darker than sepals. Median sepal hooded, egg-shaped to broadly egg-shaped, 10–15mm tall, 8–10mm deep; spur conical, 4–6mm long; lateral sepals projecting away, oblong to narrowly egg-shaped, 8–12mm long; petals with strap-shaped limb 7–10mm long, and tapering, curved or sharply bent apex; lip projecting away, oblong to elliptic, 7–12mm long, margins notched, torn or rarely entire. Highly variable in flower colour and the degree of laceration of the lip, often within populations.

| A | S | O | N | D | J | F | M | A | M | J | J |

Fairly common on well-drained, often gravelly soils, from near sea level to 1,000m. Flowers between November and January, stimulated by fire.

Disa inflexa

Syn: *Schizodium inflexum*

Mossel Bay 03.11.2008

Mossel Bay 03.11.2008

Op-die-Berg 26.09.2010

Op-die-Berg 26.09.2010

Slender, straight or flexuose terrestrial 80–350mm tall. **Basal leaves** spear-shaped, 10–20 × 5–9mm; **cauline leaves** closely sheathing the stem. **Inflorescence** 1–7-flowered. **Flowers** pink, petal apex darker pink, lip blade with darker pink to purple blotches, tooth of the lip usually dark purple. Median sepal hooded, narrowly oblong, 6–8 × 2–4mm; spur straight, rarely slightly upcurved or decurved, with apex rounded and base often widened, 5–8mm long, 2–4mm deep; lateral sepals narrowly oblong, 7.5–10 × 2–3mm; petals strap-shaped, 5–7mm long; lip 9–13mm long, with margins flat or wavy, apical tooth hanging and 3–7mm long.

Distinguishable from *D. bifida* by the smaller flowers, narrower median sepal, blunt spur seldom curved upward, later flowering time and its occurrence at higher altitudes.

A S O N D J F M A M J J

Common, but rarely forming large populations, on shallow soils above 300m, generally from 900–1,700m. Flowers between late September and November, stimulated by fire.

Disa intermedia

Motshane (Swaziland) 11.01.2010

Motshane (Swaziland) 11.01.2010

Motshane (Swaziland) 11.01.2010

Motshane (Swaziland) 11.01.2010

Fairly robust terrestrial 300–600mm tall; sterile shoot sometimes present. **Leaves** on fertile shoot 10–12, narrowly spear-shaped, up to 150mm long. **Inflorescence** lax to semi-dense, 20–30-flowered. **Flowers** cream, petals with reddish apical markings. Median sepal angled forward, hooded, 10–12mm tall; spur horizontal, gently decurved in distal half, cylindrical, 16–21mm long; lateral sepals spreading forward, oblong, 10–12mm long; petals elliptic to oblong, 10mm long; lip projecting away, linear, widening at the apex, 11mm long.

A S O N D J F M A M J J

Recorded only from Swaziland where it is localized in dry, stony grassland, from 1,300–1,500m. Flowers between January and February.

Disa introrsa

Ceres 13.12.1991

Ceres 13.12.1991

Ceres 13.12.1991

Slender terrestrial up to 150mm tall. **Leaves** 3–5, cauline, linear, sometimes with reddish margins, 50 × 1.5mm. **Inflorescence** dense, 10–20-flowered. **Flowers** non-resupinate, inwardly directed to face the floral axis, front half of median sepal, base of the petals and base of the lip dull carmine and cream. Median sepal hooded, 3mm long; spur shortly pouch-shaped, slightly ascending; lateral sepals 3.3–4mm long; petals curved, 1.9mm long; lip spoon-shaped, 2.5–3.1mm long.

A	S	O	N	D	J	F	M	A	M	J	J

Very rare in moist sandstone-derived soil, from 1,000–1,200m. Flowers between November and December, after fire.

Disa karooica

Sutherland 12.11.2013

Kamieskroon 15.11.2008

Kamieskroon 18.10.2014

Kamieskroon 15.11.2008

Slender terrestrial up to 600mm tall. **Basal leaves** strap-shaped, up to 200mm long, dry at flowering. **Inflorescence** lax, 5–15-flowered. **Flowers** cream, with purple streaks on the petals and purple dots on the galea and lip. Median sepal flat but hooded at the base, 15–25mm long; spur slender, 35–55mm long; lateral sepals 15–20mm long; petals small, strap-shaped, 8–12mm long, included in median sepal; lip narrowly spear-shaped.

Distinguishable from *D. draconis* by the lobed petal apices, and from *D. harveyana* by the petal apices that are included in the galea.

A S O N D J F M A M J J

Occasional on granite and shale soils on the semi-arid edges of the Karoo, the Kamiesberg and the Roggeveldberge, from 1,200–1,450m. Pollinated by long-proboscid flies. Flowers between October and November.

Disa klugei

Lydenburg 01.02.2011

Lydenburg 13.02.2011

Lydenburg 29.01.2014

Lydenburg 26.01.2011

Slender terrestrial 200–320mm tall. **Leaves** cauline, linear to spear-shaped, spreading, up to 90mm long. **Inflorescence** lax, 12–26-flowered. **Flowers** white, partially suffused with pale to deep pink, lateral sepals suffused with lime-yellow, spotted and blotched with purple, spur pink. Median sepal hooded, slightly flattened, 7–8 × 3–4mm; spur horizontal, decurved in the middle, 9–11mm long; lateral sepals narrowly oblong, spreading, 8 × 3mm; petals 4.5–5mm long; lip linear, projecting away and decurved near apex, 8 × 1–1.2mm.

Distinguishable from *D. patula* by the white and pink flowers and decurved spur, and from *D. vigilans* by soft leaves and a decurved, shorter spur.

A S O N D J F M A M J J

Rare and localized in rocky grassland, from 2,000–2,200m. Flowers between January and February.

Disa linderiana

Citrusdal 20.11.2006

Citrusdal 20.11.2006

De Rust 13.11.1985

Slender to stout terrestrial 90–160mm tall. **Leaves** 7–10, cauline, spear-shaped, green to beetroot-red, 75–160 × 10mm. **Inflorescence** dense, 20–25-flowered. **Flowers** facing outward to downward, white suffused with beetroot-red, median sepal with beetroot-red blotches on the inside, lateral sepals beetroot-red below. Median sepal hooded, with 2 pouches on either side of the spur; spur hanging, 3–3.3mm long; lateral sepals spreading, decurved, 6.4 × 4mm; petals 3.5–4mm long; lip spear-shaped, 5.5mm long.

| A | S | O | N | D | J | F | M | A | M | J | J |

Rare on rocky, south-facing slopes below the crest of mountain ridges, from 1,550–1,950m. Flowers between November and December.

Disa lineata

Grabouw 23.10.2009

Grabouw 24.10.2009

Grabouw 05.10.2010

Grabouw 24.10.2009

Slender, erect terrestrial 100–400mm tall. **Leaves** cauline, erect, narrowly spear-shaped, up to 100mm long. **Inflorescence** fairly dense to lax, 3–20-flowered. **Flowers** pale yellow-green, veins in sepals dark brown, lip purple with yellow base and apex, petals with purple apex. Median sepal hooded, 5mm tall, 4mm deep; spur absent; lateral sepals oblong, 6mm long; petals narrowly egg-shaped, 4mm long; lip narrowly egg-shaped, 3–4mm long.

A S O N D J F M A M J J

Very rare and occurring as isolated plants or in small populations on rock ledges in cool but often dry places in stony soil, from 500–1,700m. Flowers between September and November, after fire.

Disa longicornu

Cape Peninsula 21.12.2013

Cape Peninsula 20.12.2007

Cape Peninsula 19.12.2007

Slender, nearly flexuose lithophyte 100–200mm tall. **Basal leaves** usually 4, narrowly elliptic, stalked, 30–60mm long; **cauline leaves** mostly sheathing. **Inflorescence** single-flowered. **Flower** gunmetal blue with green veins inside galea. Median sepal hooded, apiculate, 20–25mm tall, 10–15mm deep, margins reflexed; spur cylindrical, gradually tapering, sharply decurved near apex, 20–35mm long; lateral sepals projecting away, oblong to narrowly egg-shaped, 20–30mm long; petals linear, reflexed along base of median sepal and extending to apex of the spur; lip projecting away, elliptic, 17–22mm long.

| A | S | O | N | D | J | F | M | A | M | J | J |

Localized on rock faces usually facing the southeaster clouds, from 600–1,000m. Flowers between December and January.

Disa longifolia

Op-die-Berg 08.11.1984

Op-die-Berg 08.11.1984

Citrusdal 29.10.2013

Op-die-Berg 08.11.1984

Fairly robust terrestrial 100–400mm tall. **Leaves** cauline, erect, up to 250mm long. **Inflorescence** usually dense, few- to many-flowered; bracts overtopping the flowers. **Flowers** relatively large, pink to white. Median sepal hooded, 7mm tall, 4mm deep; spur somewhat hanging and soon decurved, stout, club-shaped, 2–4mm long; lateral sepals oblong, 5–7mm long; petals curved, sharply bent midway; lip spear- to strap-shaped, 4–5mm long.

| A | S | O | N | D | J | F | M | A | M | J | J |

Rare, but occasionally forming large populations, in damp peaty soil in marshes or boggy places, from 500–1,500m. Flowers between October and November, after fire.

Disa longipetala

Syn: *Schizodium longipetalum*

Wolseley 20.08.2006

Paarl 28.08.2010

Paarl 29.08.2010

Slender, flexuose terrestrial 70–200mm tall. **Basal leaves** spatula- to strap-shaped, 8–12 × 3–6mm, margins crisped and wavy; **cauline leaves** 2 or 3, spear-shaped, 8–13mm long, upper half free. **Inflorescence** 2–7-flowered. **Flowers** greenish-yellow, lip blade and base of lateral sepals with purplish spots, thread-like extensions of petals and lip dull purple. Median sepal recurved, spear-shaped, tapering, 7–12 × 2–3mm; spur straight to slightly decurved, 2–3mm long; lateral sepals narrowly spear-shaped, recurved, 8–12 × 1.5–2mm; petals unequally bifid, with anterior tooth thread-like, at least twice as long as petal limb; lip fiddle-shaped, with apical tooth 3.5–12mm long, often reflexed.

Distinguishable from *D. biflora* by the long thread-like tooth at the end of the petal.

A S O N D J F M A M J J

Very rare and restricted to proteoid shrubland, on gravelly soils derived from Table Mountain Sandstone, from 100–550m. Flowers between August and September, after fire.

Disa lugens

Slender, reed-like terrestrial 450mm–1m tall. **Radical leaves** 8–15, rigid, between half and two thirds as long as the shoot, 2mm wide; **cauline leaves** sheathing, dry, 20–60mm long. **Inflorescence** lax, 5–25-flowered. **Flowers** variable in colour from white to creamy-green or almost black, lip usually with a striped lip blade, petals with darker vertical stripes. Median sepal hooded, narrowly egg-shaped, 12–16mm tall, 10mm deep; spur cylindrical or conical, 1–5mm long; lateral sepals projecting away, narrowly oblong, 8–13mm long; petals with linear limb, 10–15mm long, bent in the middle, distal half variable; lip egg-shaped, deeply dissected, beard-like, longer than lateral sepals, 13–19mm long.

Disa lugens var. lugens

Syn: *Herschelia lugens*

Hogsback 10.12.2008

Hogsback 30.12.2007

Hogsback 10.12.2008

Hogsback 10.12.2008

Hogsback 12.12.2013

Var. *lugens* is distinguished as follows: flowers not dark brown or maroon-black; lip usually much longer than lateral sepals.

Var. *lugens* is endemic to the Western and Eastern Cape provinces of South Africa.

Localized on coastal sand flats and dry mountain sides, in fynbos and grassland from near sea level to 1,500m. Flowers mainly between October and January, although populations in the southern Cape may flower between late April and early June.

A S O N D J F M A M J J

Disa lugens var. *nigrescens*

Humansdorp 19.01.1993

Var. *nigrescens* is distinguished as follows: flowers purplish-black; lip not much longer than lateral sepals.

    Var. *nigrescens* is endemic to the surroundings of Humansdorp in the Eastern Cape province of South Africa.

Rare and restricted to coastal plains and dunes, in sand, from 70–100m. Flowers in January.

A S O N D J F M A M J J

Disa macrostachya

Syn: *Monadenia macrostachya*

Kamieskroon 08.10.2009

Kamieskroon 11.10.2009

Kamieskroon 08.10.2009

Slender terrestrial up to 300mm tall. **Leaves** 4–8, strap-shaped, up to 130 × 20mm. **Inflorescence** lax, 10–35-flowered. **Flowers** cream, median sepal suffused with maroon, petals and lip spotted with maroon. Median sepal erect, narrowly oblong, galea 8mm tall; spur straight, somewhat inflated, 6–8mm long; lateral sepals reflexed, oblong, 7mm long; petals narrowly egg-shaped, 6mm long; lip hanging, strap-shaped, 6mm long.

A S O N D J F M A M J J

Rare amongst restios in transitional fynbos-renosterveld habitat, from 1,200–1,500m. Flowers between September and November, stimulated by fire.

Disa maculata

Porterville 13.10.2010

Cape Peninsula 26.10.2011

Cape Peninsula 15.10.2011

Cape Peninsula 28.10.2007

Slender, nearly flexuose lithophyte 60–300mm tall. **Basal leaves** forming a rosette, narrowly elliptic to narrowly spear-shaped, sometimes with a maroon edge, 10–60mm long; **cauline leaves** and bracts spotted with maroon. **Inflorescence** single-flowered. **Flower** blue, petals with purple markings. Median sepal hooded, 13–17mm tall, 7–12mm deep, narrowly egg-shaped in front view; spur absent or pouch-shaped; lateral sepals projecting away, narrowly egg-shaped to egg-shaped, often apiculate, 15–18mm long; petals linear, 7mm long; lip projecting away, narrowly spear-shaped to narrowly oblong, 15mm long.

Distinguishable from *D. virginalis*, with which it may co-occur, by the blue flower colour, the longer and narrower lateral sepals and the wide open galea entrance.

A S O N D J F M A M J J

Localized, growing in small populations in moss, on rock ledges and in crevices, from 500–1,700m. Flowers between October and November, stimulated by fire.

Disa maculomarronina

Graskop 17.01.2010

Graskop 07.01.2011

Graskop 04.01.2014

Graskop 12.01.2010

Slender to fairly robust terrestrial 300–400mm tall, producing a slender sterile shoot. **Leaves** on sterile shoot usually 2, very narrowly spear-shaped to elliptic; fertile shoot usually with 7 cauline leaves. **Inflorescence** 25–27mm wide, moderately dense, 15–35-flowered. **Flowers** from pale pink to purple-pink, petals bright yellow, lateral sepals with a maroon spot basally. Median sepal hooded, 6–7 × 3–4mm, 6.5–7.5mm deep; spur narrowly cylindrical, gently decurved, 7.5–8mm long; lateral sepals oblong to elliptic, 6–7 × 3–4mm; petals included in galea, 4–5 × 2.5–3.5mm; lip narrowly spear- to spatula-shaped, 4.2–4.5mm long.

Distinguishable from *D. versicolor* by the longer spur that is readily visible and by the flowers that do not turn brown immediately after opening.

A S O N D J F M A M J J

Rare along the edge of marshes and in seasonally flooded grasslands, from 1,500–2,300m. Flowers between December and January.

Disa marlothii

Ceres 10.02.2013

Ceres 01.02.2010

Ceres 07.02.2011

Slender terrestrial 150–350mm tall. **Leaves** cauline, lower 3–10 clustered basally, narrowly elliptic, up to 50mm long, remainder sheathing the stem. **Inflorescence** flat-topped, 1–6-flowered. **Flowers** pale pink to purplish-red, with darker spots mainly on median sepal and petals. Median sepal erect, shallowly hooded, 10–14 × 8mm; spur horizontal, cylindrical, often somewhat curved upward at apex, 15–20mm long; lateral sepals projecting away, oblong, 10–15mm long; petals narrowly oblong, 8mm long; lip projecting away, narrowly elliptic, 10mm long.

| A | S | O | N | D | J | F | M | A | M | J | J |

Rare in rock crevices beside streams, from 600–1,300m. Flowers between December and February, stimulated by fire.

Disa micropetala

Grabouw 30.10.2009

Grabouw 25.10.2009

Grabouw 30.10.2009

Slender terrestrial 60–300mm tall. **Leaves** numerous, cauline, linear, up to 120mm long. **Inflorescence** dense, many-flowered. **Flowers** minute, with median sepal white mottled with pale pink, lateral sepals and lip very dark purple to beetroot-red from a white base. Median sepal almost spherical, 2–4mm in diameter; spur hanging and decurved, 1mm long; lateral sepals 2–4mm long; petals 2mm long; lip strap-shaped, fleshy, 2.5mm long.

A S O N D J F M A M J J

Local and rare in damp peaty soils and on rock ledges, from 800–1,650m. Flowers between the middle of October and December, after fire.

Disa minor

Op-die-Berg 16.11.2009

Slender terrestrial or lithophyte 50–100mm tall. **Basal leaves** very narrowly elliptic to elliptic, 10–15mm long. **Inflorescence** flat-topped, 3–6-flowered. **Flowers** non-resupinate, pink. Median sepal shallowly hooded, rounded, 6mm tall; spur rudimentary; lateral sepals egg-shaped, apiculate, 7mm long; petals narrowly oblong, 3mm long; lip strap- to narrowly spear-shaped, 4mm long.

A S O N D J F M A M J J

Very rare and localized in moss on wet rock ledges or in marshes on high mountains, from 1,200–2,000m. Flowers between November and December, after fire.

Disa montana

Maclear 18.12.2010

Maclear 20.12.2010

Maclear 18.12.2010

Maclear 18.12.2010

Slender terrestrial 300–600mm tall. **Leaves** cauline, rigid, linear to spear-shaped, up to 120mm long. **Inflorescence** lax, 10–20-flowered. **Flowers** white to deep pink with few or many maroon spots, variable even within a population. Median sepal shallowly hooded, 10–18mm tall; spur horizontal, cylindrical, 8–15mm long; lateral sepals projecting away, oblong, 12–17mm long, apiculi pointed; petals narrowly spear-shaped, 6–9mm long; lip projecting away, decurved near the tip, narrowly oblong, 10–15mm long.

A S O N D J F M A M J J

Rare on dry rocky grassland slopes in full sun, from 1,000–2,400m. Flowers between November and December.

Disa multifida

Syn: *Herschelia multifida*

Ceres 02.12.2008

Ceres 30.11.2007

Ceres 30.11.2007

Ceres 30.11.2007

Slender, reed-like terrestrial 400–600mm tall. **Radical leaves** 10–20, rigid, usually about half as long as inflorescence, 1mm wide; **cauline leaves** completely sheathing, 20–50mm long. **Inflorescence** lax, 3–8-flowered. **Flowers** blue with green lip, often with greenish veins and brownish spur. Median sepal hooded, egg-shaped, 10–20mm tall, 6mm deep; spur conical, 1–6mm long; lateral sepals projecting away, spear-shaped to narrowly egg-shaped, 10–15mm long; petals with curved limb, linear, 7–10mm long, apex tapering or narrowly triangular, occasionally torn; lip with a linear 30–100mm-long limb, blade narrowly egg-shaped, deeply torn.

| A | S | O | N | D | J | F | M | A | M | J | J |

Localized on dry stony mountain slopes, occasional in slightly moister conditions, from 200–2,000m. Flowers between November and December.

Disa neglecta

Oudtshoorn 14.12.2006

Oudtshoorn 14.12.2006

Oudtshoorn 21.12.2005

Slender terrestrial 90–180mm tall. **Leaves** 5–8, cauline, linear, up to 90mm tall, reaching above inflorescence base. **Inflorescence** dense, many-flowered. **Flowers** facing downward, dull cream or green to purple-red, median sepal pure cream-green or variously spotted with red-brown. Median sepal hooded, 4mm tall, 2mm deep, with 2 rounded basal pouches; spur very short or absent; lateral sepals 4mm long; petals 1mm long; lip linear, 3mm long.

Distinguishable from *D. obtusa* subsp. *picta* by the spur less than 0.5mm long and appearing bifid.

A S O N D J F M A M J J

Rare in drier, rocky, burnt fynbos habitat, from 1,150–1,900m. Flowers between November and December, after fire.

Disa nervosa

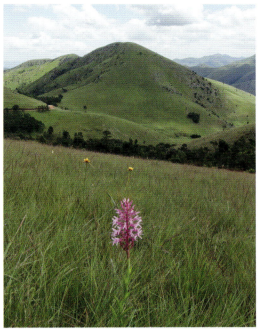

Josefsdal 03.02.2011

Oshoek 29.01.2012

Ngeli 18.02.2008

Ngeli 18.02.2008

Robust terrestrial 400–800mm tall. **Leaves** 8–11, narrowly spear-shaped, up to 250mm long. **Inflorescence** fairly dense, 20–50-flowered. **Flowers** pink with purple markings or spots at petal apices. Median sepal narrowly oblong, shallowly hooded, 15–25mm long; spur horizontal, gently decurved near the tip, thread-like, 12–20mm long; lateral sepals strap-shaped to narrowly spear-shaped, 15–25mm long; petals strap-shaped, mostly included in galea, 13–23mm long; lip projecting away, linear, often somewhat widened apically, 17–28mm long.

A S O N D J F M A M J J

Widespread and occasional in dry, usually stony, high-altitude grassland, in full sun, from 300–2,000m. A mimic of *Watsonia densiflora* flowers. Pollinated by long-proboscid flies. Flowers between January and February.

Disa nivea

Sani Pass 28.01.2006

Ramatseliso 24.02.2007

Sani Pass 17.01.2014

Elliot 25.02.2010

Slender to fairly robust terrestrial 200–400mm tall. **Leaves** up to 10, cauline, rigid, erect, linear, up to 250mm long. **Inflorescence** semi-lax, 10–20-flowered. **Flowers** white or cream, often with red spots at lateral sepal apices. Median sepal hooded, 9–12 × 6mm, 4mm deep; spur cylindrical, horizontal and decurved near the tip, 25–40mm long; lateral sepals oblong, shortly apiculate, 9–13mm long; petals included in galea, narrowly oblong, 5mm tall; lip projecting away, spear-shaped, 10mm long.

A S O N D J F M A M J J

Occasional on rock ledges or steep rocky slopes, in full sun, from 1,700–2,500m. A mimic of *Zaluzianskya* flowers. Pollinated by long-proboscid flies. Flowers between January and February.

Disa nubigena

Cape Peninsula 21.11.1998

Worcester 06.12.2008

Cape Peninsula 21.11.1998

Slender terrestrial 140–240mm tall. **Leaves** narrowly spear-shaped, up to 100 × 40mm. **Inflorescence** loosely cylindrical, up to 60mm long, 10–45-flowered. **Flowers** 10–12mm in diameter, sepals white, lip and petals yellow to brown apically. Median sepal shallowly hooded, curved forward, 5mm long; spur tapering, 2mm long; lateral sepals spreading, spear-shaped, 4.5mm long; petals square, 2.5 × 2mm, shallowly 2-lobed, front lobe combining with median sepal in forming a deep galea; lip linear, 4mm long.

A S **O** N **D** J F M A M J J

Very rare and known only from two localities in mature restioid fynbos, from 1,000–1,800m. Flowers between October and December.

Disa obliqua

Slender, flexuose terrestrial 55–290mm tall. **Basal leaves** broadly egg- to spatula-shaped, 5–14 × 3–9mm; **cauline leaves** spear-shaped, sheathing, 7–18mm long. **Inflorescence** 2–9-flowered. **Flowers** pale pink, lip with a dotted pattern, apical tooth dark-coloured. Median sepal almost erect, spear-shaped, 5–10 × 1.5–4mm; spur linear to club-shaped, constricted at the base, 3–9mm long; lateral sepals spear-shaped, 5–14 × 1.5–4mm; petals included in galea, apices variously bifid; lip with apical wedge- to awl-shaped tooth.

Disa obliqua subsp. *clavigera*

Syn: *Schizodium obliquum* subsp. *clavigerum*

Rooiels 28.08.2007

Rooiels 12.09.2008

Rooiels 11.09.2008

Rooiels 12.09.2008

Subsp. *clavigera* is distinguished as follows: lateral sepals generally shorter than 8.5mm, about as long as median sepal; plants 80–240mm tall.

Subsp. *clavigera* is endemic to the more inland areas of the south-western corner of the Western Cape from Bredasdorp in the east to Ceres in the west.

Occasional on well-drained sandy to stony soil on hills or mountain slopes, from near sea level to 600m. Flowers between late August and September, stimulated by fire.

| A | S | O | N | D | J | F | M | A | M | J | J |

Disa obliqua subsp. *obliqua*

Syn: *Schizodium obliquum* subsp. *obliquum*

Cape Peninsula 28.08.2010

Villiersdorp 24.08.2010

Villiersdorp 23.08.2010

Houwhoek 27.08.2010

Subsp. *obliqua* is distinguished as follows: lateral sepals generally longer than 8.5mm, about one third longer than median sepal; plants 100–250mm tall.

Subsp. *obliqua* is endemic to the coastal plains of the south-western corner of the Western Cape, mainly from the Cape Peninsula to Cape Hangklip and around Stellenbosch, Malmesbury and Darling.

Occasional on seasonally damp sand, from near sea level to 500m. Flowers mainly between August and September, stimulated by fire.

A S O N D J F M A M J J

Disa obtusa

Slender to moderately robust terrestrial 60–400mm tall. **Leaves** cauline, linear, 30–120mm long. **Inflorescence** slender, dense, many-flowered. **Flowers** facing downward, white to brown or purple, often finely spotted. Median sepal 3–8mm tall, galea almost round; spur somewhat hanging, with a deep dorsal groove extending up the galea, 0.2–2.2mm long; lateral sepals spreading, 3–6mm long; petals curved, 2mm long; lip strap-shaped, 2.5–5mm long.

Disa obtusa subsp. *hottentotica*

Grabouw 27.11.2009   Grabouw 27.11.2009

Grabouw 14.10.2008

Villiersdorp 18.10.2008   Villiersdorp 18.10.2008

Subsp. *hottentotica* is distinguished as follows: median sepal 5–8mm long; spur straight, 0.2–1.5mm long.

Subsp. *hottentotica* is restricted to the Hottentots Holland Mountains in the Western Cape, from Jonkershoek to Betty's Bay.

Localized but often common on deep peaty soils, in seeps, marshes and on dryish slopes, from near sea level to 1,500m. Hybridizes with *D. albomagentea*. Flowers between September and December, after fire.

A S O N D J F M A M J J

Disa obtusa subsp. *obtusa*

Cape Peninsula 14.10.2007

Cape Peninsula 05.10.2007

Cape Peninsula 06.10.2008

Cape Peninsula 04.12.2013

Cape Peninsula 06.10.2008

Subsp. *obtusa* is distinguished as follows: flowers white or brown with brownish mottling; median sepal 4.5–5.5mm long; spur straight, 0.5–1mm long.

Subsp. *obtusa* is restricted to the Cape Peninsula of the Western Cape.

Localized but often common in deep sandy soils, often in damp places, from sea level to 1,000m. Pollinated by March flies. Flowers between September and December, after fire.

A	S	O	N	D	J	F	M	A	M	J	J

Disa obtusa subsp. *picta*

Grabouw 31.10.2009

Grabouw 25.10.2009

Grabouw 24.10.2009

Subsp. *picta* is distinguished as follows: flowers purplish mottled on a white base; median sepal 3–5mm long; spur sharply decurved, 1–2mm long.

Subsp. *picta* occurs in the western and southern parts of the Cape Floral Region from the Hottentots Holland Mountains to Humansdorp.

Distinguishable from *D. neglecta* by the spur usually longer than 0.4mm and not bifid.

Widespread and locally common on peaty, south-facing slopes, often in slight seepages, from 300–1,500m. Flowers between October and December, after fire.

A S O N D J F M A M J J

Disa ocellata

Villiersdorp 29.11.2009

Villiersdorp 29.11.2009

Cape Peninsula 19.12.2007

Cape Peninsula 19.12.2007

Slender, erect terrestrial 50–300mm tall. **Leaves** cauline, erect, linear, up to 80 × 8mm. **Inflorescence** lax, up to 20-flowered; bracts usually overtopping the flowers. **Flowers** white or purplish-pink, with brownish veins and a dull purple-brown patch on either side of the galea, the underside of the lateral sepals and the apex of the spur. Median sepal hooded, 4mm tall, 5mm deep; spur horizontal, club-shaped in side view, usually sharply notched apically, 2mm long; lateral sepals oblong, 5mm long; petals erect, 4mm long; lip strap-shaped, 4mm long.

Distinguishable from *D. uncinata* by the brown patches on the galea.

A	S	O	N	D	J	F	M	A	M	J	J

Rare on mountain summits on moss banks or stony soils in the zone of the southeaster clouds, from 900–1,900m. Flowers between October and December, mainly within two years after fire.

Disa oligantha

Grabouw 02.01.2010

Grabouw 02.01.2010

Grabouw 02.01.2010

Cape Peninsula 07.12.1990

Dwarf terrestrial 40–150mm tall. **Basal leaves** usually 6, narrowly spear-shaped, up to 15mm long; **cauline leaves** sheathing at base, free at tip, linear to spear-shaped. **Inflorescence** flat-topped, up to 10-flowered. **Flowers** non-resupinate, sepals cream, petals and lip yellow from white bases. Median sepal hooded, 3-lobed, 4–5mm tall; spur absent; lateral sepals oblong, concave, 6–7mm long; petals oblong; lip projecting away, linear, 2mm long.

A S O N **D J** F M A M J J

Rare on mountain summits in stony soil in the southeaster cloud zone, from 950–1,400m. Flowers between December and January, after fire.

Disa ophrydea

Syn: *Monadenia ophrydea*

Grabouw 06.10.2010

Cape Peninsula 05.10.2007

Cape Peninsula 14.10.2007

Grabouw 06.10.2010

Grabouw 30.10.2009

Cape Peninsula 11.10.2009

Slender terrestrial up to 400mm tall, suffused with beetroot-red. **Leaves** green, narrowly spear-shaped, basally up to 100 × 20mm. **Inflorescence** lax, 5–15-flowered; bracts purplish-green. **Flowers** purple-red, lateral sepals paler and often yellow or white. Median sepal oblong to egg-shaped, shallowly hooded, 9–11 × 5–7mm; spur slender, tapering, 20–24mm long; lateral sepals spreading, narrowly egg-shaped, 8–10mm long; petals very obliquely egg-shaped; lip narrowly elliptic to almost strap-shaped, fleshy, 8–10mm long.

Distinguishable from *D. atrorubens* by the spreading lateral sepals and narrow lip.

A S O N D J F M A M J J

Fairly common and usually in extensive populations in damp conditions, from 300–1,000m. Pollinated by small moths. Flowers between October and November, after fire.

Disa oreophila

Slender erect or flexuose terrestrial 100–350mm tall. **Leaves** linear, rigid, sharply pointed, up to 200 × 1–3mm. **Inflorescence** lax, 15–25-flowered. **Flowers** white or pink with purple spots on sepals. Median sepal hooded, 5–6 × 4mm, 2mm deep, margins not bending inward; spur cylindrical, straight or gradually decurved, 5–10mm long; lateral sepals projecting away, oblong, 4–7mm long, apiculi well-developed; petals narrowly oblong, 2.5–4mm tall; lip spear-shaped, 4–6mm long.

Disa oreophila subsp. *erecta*

Naudesnek 07.01.2009

Naudesnek 05.02.2008

Naudesnek 05.02.2008

Subsp. *erecta* is distinguished as follows: habit erect, stout; median sepal 6–10mm long; spur 10–25mm long.

Subsp. *erecta* is endemic to the Drakensberg of South Africa and Lesotho.

Occasional on rock ledges and damp grassy slopes, from 2,250–2,700m. Pollinated by long-proboscid flies. Flowers between January and February.

| A | S | O | N | D | J | F | M | A | M | J | J |

Disa oreophila subsp. *oreophila*

Garden Castle 16.01.2014

Luneburg 07.01.2014

Garden Castle 16.01.2014

Garden Castle 16.01.2014

Subsp. *oreophila* is distinguished as follows: habit flexuose; median sepal 4–6mm long; spur 5–10mm long.

Subsp. *oreophila* is endemic to the Drakensberg of South Africa and Lesotho.

Distinguishable from *D. nivea* by the smaller flowers and shorter spur, and from *D. saxicola* by the much narrower, rigid leaves.

Occasional in rock crevices and on ledges, from 1,200–2,100m. Flowers between December and February, mainly in January.

| A | S | O | N | D | J | F | M | A | M | J | J |

Disa ovalifolia

Op-die-Berg 25.08.2013

Op-die-Berg 26.09.2012

Op-die-Berg 06.08.2011

Op-die-Berg 25.09.2012

Slender to fairly robust terrestrial 120–200mm tall. **Leaves** 8–10, cauline, lower 4–6 egg-shaped, often maroon on underside. **Inflorescence** dense, 10–15-flowered. **Flowers** facing downward, sepals green, lip brown. Median sepal hooded, 7–9mm tall, 2–3mm deep; spur somewhat ascending from a conical base, straight or curved near apex, 10–15mm long; lateral sepals spreading sideways, oblong, 8mm long; petals narrowly oblong, 7mm long; lip hanging, strap-shaped, rather fleshy, 8–10mm long.

A S O N D J F M A M J J

Rare in dry, sandy soils in mature veld, from 400–1,300m. Flowers between July and September, stimulated by fire.

Disa patula

Slender to fairly robust terrestrial 250–600mm tall. **Leaves** cauline, linear to narrowly spear-shaped, up to 200mm long. **Inflorescence** lax, 14–50-flowered. **Flowers** pale pink to pink, petals and inner surface of galea with small purple spots. Median sepal hooded, 6–10 × 2.5mm, 2mm deep; spur horizontal, cylindrical, tapering, 5–12mm long; lateral sepals spreading forward, narrowly oblong, 6–10mm long; petals narrowly egg- to spear-shaped, 5–10mm long; lip projecting away, linear, somewhat widened apically, 9mm long.

Disa patula var. patula

Sabie 25.01.2014

Sabie 25.01.2014

Sabie 25.01.2014

Var. *patula* is distinguished as follows: inflorescence generally wider than 30mm, usually 40–50mm; median sepal longer than 10mm or, if shorter, then spur shorter than 10mm.

Var. *patula* is endemic to South Africa.

Rare in grassland, from 1,500–2,000m. Flowers between November and January.

| A | S | O | N | D | J | F | M | A | M | J | J |

Disa patula var. *transvaalensis*

Giant's Castle 03.02.2009

Ramatseliso 13.12.2008

Sabie 29.01.2014

Dullstroom 01.02.2011

Var. *transvaalensis* is distinguished as follows: inflorescence slender, narrower than 35mm; median sepal shorter than 10mm and spur longer than 8mm.

Var. *transvaalensis* occurs in South Africa, Swaziland and Zimbabwe.

Widespread and locally common in dry to slightly damp grassland, from 600–2,100m. Flowers between December and January.

Disa perplexa

Luneburg 16.01.2012

Luneburg 22.01.2014

Pietermaritzburg 18.12.2012

Luneburg 11.01.2013

Luneburg 18.01.2012

Slender terrestrial 260–600mm tall. **Leaves** cauline, spear-shaped, 40–80 × 10–15mm. **Inflorescence** fairly dense, 10–15-flowered. **Flowers** facing downward, greenish-white, except for petals and lip variously speckled and mottled with maroon to purple. Median sepal hooded, 5.5–6.5 × 4–4.5mm, 2mm deep; spur horizontal from a conical base, abruptly decurved in the middle, 9–11mm long; lateral sepals projecting away, 5–5.5 × 3–3.5mm; petals included in galea; lip egg-shaped, 3–5 × 1.3–2mm.

Similar to *D. hircicornis*, which has been recorded for South Africa but has not been seen for a long time. The flower of *D. hircicornis* is uniformly purple and has a narrowly oblong lip, less than 1.3mm wide.

Rare in marshy grassland, from 900–2,200m. Flowers between December and January.

Disa physodes

Tulbagh 19.09.2011

Paarl 17.10.2009

Paarl 17.10.2009

Tulbagh 19.09.2011

Ceres 09.10.2009

Robust terrestrial 250–600mm tall. **Leaves** linear to spear-shaped, longest at the base, 140–200 × 20mm. **Inflorescence** fairly dense, many-flowered. **Flowers** lime- to yellow-green, sepals often mottled or suffused with maroon. Median sepal shallowly hooded, oblong, 9–11mm tall; spur club-shaped, 7–9 × 3–6mm, pressed close to the ovary; lateral sepals reflexed, oblong, 7–10mm long; petals unevenly egg-shaped to oblong, 7–9mm long; lip strap- to narrowly spear-shaped, 7–10mm long.

Distinguishable from *D. cernua* by the shorter spur, less than 10mm long.

| A | S | O | N | D | J | F | M | A | M | J | J |

Rare on clay soils in renosterveld or renosterveld-fynbos transition, from near sea level to 1,000m. Flowers between September and October, after fire.

Disa pillansii

Kleinmond 22.11.2010

Stanford 15.10.2013

Stanford 15.10.2013

Stanford 15.10.2013

Slender terrestrial or lithophyte 150–300mm tall. **Basal leaves** 3–6, elliptic, 30–60mm long. **Inflorescence** almost flat-topped, 2–15-flowered. **Flowers** purplish-pink, petals lilac below and yellow barred with purple above. Median sepal 8–11mm tall, galea almost spherical, 6mm wide, 5mm deep, entrance reduced to a narrow slit; spur rudimentary; lateral sepals spreading upward, egg-shaped, deeply concave, 10–14mm long; petals strap-shaped, curved, 5–6mm long; lip linear to strap-shaped, 4–6mm long.

A S O N D J F M A M J J

Rare but often locally abundant on stream banks and mossy seepages on cliff faces in the southeaster cloud zone, from 300–1,500m. Pollinated by small bees. Flowers between October and December, stimulated by fire.

Disa polygonoides

KZN south coast 11.12.2011

KZN south coast 11.01.2011

KZN south coast 28.01.2010

KZN south coast 28.01.2010

Slender to robust terrestrial 150–700mm tall, rarely producing a sterile shoot. **Leaves** of fertile shoot 6–25, linear to strap-shaped, up to 250 × 40mm (far fewer and smaller if a sterile shoot is present). **Inflorescence** dense, cylindrical, 40–100-flowered. **Flowers** brick- to orange-red, less than 10mm in diameter. Sepals 5–7mm long; median sepal erect; spur abruptly hanging and reaching below base of median sepal, 3–5mm long; petals curved over anther; lip strap-shaped to linear, 6 × 1mm.

Distinguishable from *D. woodii* by the red flowers and longer spur reaching just below the base of the median sepal, and from *D. chrysostachya* by the shorter, non-inflated spur.

| A | S | O | N | D | J | F | M | A | M | J | J |

Fairly common along the eastern coastal areas, with some inland outliers, in marshy or grassland habitats, from near sea level to 1,500m. Flowers between October and May, and occasionally also in winter.

Disa porrecta

Misgund 24.03.2011   Rhodes 25.01.2010   Rhodes 25.01.2010

Misgund 24.03.2011   Rhodes 25.01.2010   De Rust 22.03.2011

Fairly stout, reed-like terrestrial 200–600mm tall. **Radical leaves** developing after flowering, linear, up to 300mm long, often persisting as a fibrous sheath; **cauline leaves** sheathing. **Inflorescence** dense, 5–15-flowered. **Flowers** facing downward at 45°, pink, orange, bright red or scarlet, petals and lip yellow. Median sepal deeply hooded, 5 × 5mm; spur ascending, 20–40mm long; lateral sepals broadly egg-shaped, 6–8mm long, apiculi up to 3mm long; petals spear-shaped to oblong, 4mm tall; lip spear-shaped to narrowly oblong, 7–10mm long.

A S O N D J F M A M J J

Widespread in grassland or fynbos, from 600–2,000m. Pollinated by the mountain pride butterfly and long-proboscid flies. Flowers between January and March, stimulated by fire.

Disa procera

Wilderness 18.10.2011

Wilderness 03.11.2012

Wilderness 05.11.2008

Slender, reed-like terrestrial 400–600mm tall. **Radical leaves** 8–13, green at flowering time, semi-rigid, up to 2mm wide; **cauline leaves** completely sheathing, 20–40mm long. **Inflorescence** lax, 6–30-flowered. **Flowers** deep cerise-red. Median sepal hooded, egg-shaped to broadly egg-shaped, 9–12mm tall; spur conical, 2–3mm long; lateral sepals projecting away, oblong, 8–10mm long; petals with strap-shaped limb, 7–10mm long, curved or bent like a knee, more or less 2-lobed apically; lip projecting away, oblong, margins notched with small, regular teeth to torn, rarely entire, 8–10mm long.

Distinguishable from *D. hians* by the cerise flowers and shorter spur.

A S O N D J F M A M J J

Very rare near Wilderness, from 30–40m. Flowers between October and November.

Disa pulchra

Witsieshoek 17.12.2008

Witsieshoek 17.12.2008

Greytown 03.01.2011

Greytown 31.12.2009

Witsieshoek 17.12.2008

Fairly robust to slender terrestrial 250–600mm tall. **Leaves** rigid, linear to spear-shaped, up to 200mm long. **Inflorescence** fairly dense, rarely lax, 5–20-flowered. **Flowers** rose-pink, parts spreading forward to form a pseudo-tube. Median sepal angled forward, narrowly egg-shaped, shallowly hooded, 20–30mm long; spur horizontal, apically decurved, cylindrical, 10–20mm long; lateral sepals spreading forward with somewhat reflexed apices, narrowly oblong to oblong, 20–30mm long; petals spear-shaped, 10–15mm long; lip projecting away, spear-shaped, 20–30mm long.

A S O N **D** J F M A M J J

Occasional in stony grassland in hilly or mountainous country, from 1,200–2,100m. Mimic of *Watsonia lepida*. Pollinated by long-proboscid flies. Flowers in December.

Disa purpurascens

Syn: *Herschelia purpurascens*

Cape Peninsula 24.10.2006   Betty's Bay 08.11.2008   Cape Peninsula 09.11.2007

Cape Peninsula 23.11.2013   Cape Peninsula 23.11.2013

Slender, reed-like terrestrial 250–500mm tall. **Radical leaves** about 10, rigid, erect, up to 1mm wide; **cauline leaves** 5–7, completely sheathing, 20–40mm long. **Inflorescence** lax, 1–7-flowered. **Flowers** blue, lip more purplish than sepals, rear lobes of petals yellow or green. Median sepal hooded, egg-shaped, 15–25mm tall, 10–15mm deep; spur conical, horizontal or slightly curved upward, 1–4mm long; lateral sepals projecting away, oblong, 15–18mm long; petals with strap-shaped limb 8–10mm long, apex expanded into a 4–5mm-broad fan notched with teeth at the margin; lip broadly egg-shaped, with a short claw, margins crisped and curved upward, 12–18mm long.

A S O N D J F M A M J J

Distinguishable from *D. graminifolia* by the conical spur, upcurved lip margins and earlier flowering time.

Uncommon in well-drained, rocky or stony areas, from near sea level to 300m. Flowers between October and November.

Disa pygmaea

Syn: *Monadenia pygmaea*

Kleinmond 06.10.2011

Kleinmond 08.10.2011

Betty's Bay 28.10.2008

Kleinmond 09.10.2011

Kleinmond 09.10.2011

Dwarf terrestrial 45–150mm tall. **Leaves** cauline, narrowly egg-shaped, 15–20mm long. **Inflorescence** cylindrical, dense, 15–100mm long, longer than leafy part of stem. **Flowers** with lip and petals lime-green, lateral sepals green, median sepal and spur rusty brown. Median sepal shallowly hooded, oblong, 5–6mm long; spur curved towards ovary, 2.5mm long; lateral sepals projecting away, apices reflexed, oblong, 4mm long; petals shortly and acutely bifid apically, 4mm long; lip somewhat hanging, strap-shaped, 3–4mm long.

| A | S | O | N | D | J | F | M | A | M | J | J |

Rare and in small populations in sandy areas, from near sea level to 1,000m. Flowers between September and November, after fire.

Disa racemosa

Cape Peninsula 21.12.2007

Cape Peninsula 21.12.2007

Houwhoek 26.11.2010

Cape Peninsula 25.12.2007

Cape Peninsula 27.11.2007

Slender terrestrial up to 1m tall. **Basal leaves** 3–10, narrowly spear-shaped, up to 100mm long. **Inflorescence** lax, 2–8-flowered. **Flowers** pink or very pale pink, veins darker, petals and lip white to yellow, petals with horizontal purple bars. Median sepal shallowly hooded, 17–24 × 13–20mm, 5–8mm deep; spur barely visible; lateral sepals projecting away, oblong to broadly elliptic, 15–25mm long, apiculi 1mm long; petals narrowly egg-shaped to quadrangular, 10–15mm long; lip projecting away, linear, 10mm long.

Distinguishable from *D. venosa* by the broadly elliptical to broadly egg-shaped median sepal.

A	S	O	N	D	J	F	M	A	M	J	J

Widespread and often common in swampy areas, from near sea level to 2,000m. Pollinated by carpenter bees. Flowers between October and December, after fire.

Disa remota

Worcester 09.12.2003

Worcester 07.12.2003

Slender terrestrial up to 150mm tall. **Leaves** 4, cauline, linear to spear-shaped, suffused with beetroot-red, 65 × 10mm. **Inflorescence** moderately dense, 18–20-flowered; lowermost bracts overtopping the flowers. **Flowers** white with purple mottling, spur green. Median sepal shallowly hooded, 9 × 5mm; spur hanging, straight, 2.5mm long; petals 3.5 × 0.9mm; lip 3-lobed; lip midlobe linear, 7.8mm long.

A S O N D J F M A M J J

Very rare on rocks on steep north-facing slopes at about 1,700m. Flowers between November and December.

Disa reticulata

Syn: *Monadenia reticulata*

Betty's Bay 01.11.2008

George 02.12.2007

Joubertina 14.11.2010

Betty's Bay 14.11.2008

George 13.12.2010

Betty's Bay 02.11.2008

Slender terrestrial 80–400mm tall. **Leaves** linear to spear-shaped, up to 150mm long. **Inflorescence** dense, cylindrical, many-flowered; bracts visibly net-veined. **Flowers** lime-green, petals or sepals occasionally with some maroon tinting or mottling. Median sepal shallowly hooded, oblong, 7–8mm long; spur slender, longer than galea, 10–20mm long; lateral sepals oblong, 6–7mm long; petals narrowly egg-shaped, 5–6mm long; lip hanging, strap-shaped, 4–6.5mm long.

| A | S | O | N | D | J | F | M | A | M | J | J |

Occasional in seasonally damp places, from 200–1,500m. Flowers between October and December, after fire.

Disa rhodantha

Dullstroom 14.01.2010

Dullstroom 14.01.2010

Dullstroom 30.01.2012

Slender to moderately robust terrestrial 300–600mm tall, producing a slender sterile shoot. **Leaves** of sterile shoot 3 or 4; fertile shoot with 8–16 cauline leaves. **Inflorescence** dense, 20–50-flowered. **Flowers** facing downward, 15mm in diameter, pink, middle and front of galea often lighter. Median sepal hooded, facing downward, 7–10 × 4–5mm, 3.5–4.5mm deep; spur slender, cylindrical, upcurved and ascending, 7–17mm long; lateral sepals spreading horizontally, oblong, 7–10 × 3.5–5mm, apiculi rudimentary to 1.5mm long; petals erect, 5.5–8 × 1–2mm; lip hanging, spear-shaped, 6–7 × 1–2mm.

A S O N D J F M A M J J

Localized in swamps, on stream banks or in wet grassland, from 1,500–2,100m. Pollinated by long-proboscid flies. Hybridizes rarely with *D. chrysostachya*. Flowers between late December and February.

Disa richardiana

Cape Peninsula 31.10.2008

Cape Peninsula 22.10.2009

Cape Peninsula 22.10.2009

Cape Peninsula 22.10.2009

Slender terrestrial or lithophyte 40–300mm tall. **Basal leaves** up to 10, elliptic, up to 35mm long. **Inflorescence** flat-topped, 3–8-flowered. **Flowers** non-resupinate, white, petals and lip yellow. Median sepal hooded, 7–10 × 4mm, 5mm deep, margins bending slightly inward; spur rudimentary; lateral sepals spreading upward, egg-shaped, deeply concave, 9–12mm tall; petals triangular, 5–7mm long; lip projecting away, spear- to egg-shaped, 3–6mm long.

A S O N D J F M A M J J

Occasional in moss on rock ledges or damp turfy areas in the southeaster cloud zone, from 700–1,500m. Flowers between September and November, stimulated by fire.

Disa rosea

Cape Peninsula 31.10.2008

Cape Peninsula 31.10.2008

Cape Peninsula 31.10.2008

Cape Peninsula 22.10.2009

Slender, flexuose lithophyte 60–200mm tall. **Basal leaves** 2 or 3, egg-shaped, 25–70mm long, purple below. **Inflorescence** lax to dense, almost flat-topped, arching horizontally, 1–12-flowered. **Flowers** pale pink to almost pure white. Median sepal erect, 8–13mm tall, 4mm deep; spur rudimentary; lateral sepals projecting away, elliptic to oblong, apiculate, 8–13mm long; petals almost triangular, 5 × 2.5mm; lip projecting away, narrowly oblong, somewhat widened basally, 5–7mm long.

A S O N D J F M A M J J

Occasional on mossy rock ledges in seepages in partial shade of rocky overhangs in the southeaster cloud zone, from 600–1,000m. Self-pollinating. Flowers between late October and December.

Disa roseovittata

Luneburg 21.01.2010

Wakkerstroom 10.01.2011

Lochiel 08.01.2014

Lochiel 10.01.2013

Robust terrestrial 520–830mm tall. **Leaves** cauline, up to 12, up to 150 × 33mm. **Inflorescence** fairly dense, 30–50-flowered. **Flowers** white suffused with pink, petals with a single dark pink stripe, lip white to pale pink. Median sepal hooded, 25–27 × 5–7mm; lateral sepals spreading, apiculate, 21–27 × 5–7mm; petals strongly curved, well exserted from galea, 19–21 × 6–7mm; lip linear, gently recurved, 20–23 × 1–2mm.

Distinguishable from *D. nervosa* by its more robust habit and strongly curved petals that are exserted from the galea and marked with a single stripe.

A S O N D J F M A M J J

Rare and localized in high-altitude, rocky grassland, from 1,000–2,000m. Flowers between December and January.

Disa rufescens

Syn: *Monadenia rufescens*

Cape Peninsula 23.09.2007

Cape Peninsula 29.08.2010

Cape Peninsula 23.09.2007

Cape Peninsula 29.08.2010

Slender terrestrial 140–400mm tall. **Leaves** cauline, narrowly spear-shaped to linear, lowest leaves up to 70mm long. **Inflorescence** lax, 2–25-flowered; bracts dark green and partially obscuring the flowers. **Flowers** lime-green, petals and lip dark purple. Median sepal shallowly hooded, oblong, 9–11mm long; spur cylindrical, 10–16mm long; lateral sepals oblong, 6–8mm long; petals unevenly oblong, 6–7mm long, narrowly oblong apical part forming a tube with median sepal; lip narrowly oblong, 6–7mm long.

A S O N D J F M A M J J

Occasional in seasonally damp places, from near sea level to 1,500m. Flowers between August and November, with most records from September and October.

Disa rungweensis

Sabie 01.02.2011

Sabie 01.02.2011

Sabie 24.01.2014

Sabie 25.01.2014

Sabie 24.01.2014

Dwarf terrestrial 150–250mm tall. **Basal leaves** 6 or 7, narrowly elliptic to narrowly egg-shaped, margins wavy, 45–70 × 6–7mm; **cauline leaves** usually 3. **Inflorescence** dense, 25–40-flowered. **Flowers** facing downward, greenish, suffused with red-brown. Median sepal deeply hooded, almost hemispherical, 2.5–3 × 1.5–3mm; spur distinctive, ascending, apex hammer-like, 3–4mm long; lateral sepals clasping median sepal on the sides, oblong to egg-shaped, concave, 2.5 × 1.5mm; petals mostly included in galea; lip projecting forward, narrowly oblong, 1.3–1.5mm.

| A | S | O | N | D | J | F | M | A | M | J | J |

Rare in shallow soil over bedrock, from 1,900–2,300m. Flowers between January and February.

Disa sabulosa

Syn: *Monadenia sabulosa*

Betty's Bay 09.10.2008

Betty's Bay 30.10.2008

Betty's Bay 09.10.2008

Betty's Bay 16.10.2008

Betty's Bay 18.10.2008

Fairly stout terrestrial 80–100mm tall. **Leaves** cauline, spear-shaped, margins wavy, lowest leaves up to 50mm long. **Inflorescence** fairly dense, many-flowered. **Flowers** with sepals lime-green, petals yellow. Median sepal almost spatula- to broadly egg-shaped, shallowly hooded, 10–15mm long; spur hanging, flexuose near the base, 10–15mm long; lateral sepals reflexed, oblong, 7–9mm long; petals deeply and equally bifid apically, twisted to face forward, 7–8mm long; lip spear-shaped, hanging, 6mm long.

A S O N D J F M A M J J

Rare on damp to dry sand, from near sea level to about 100m. Hybridizes rarely with *D. atrorubens*. Flowers between September and November, after fire.

Disa sagittalis

Wilderness 17.10.2011

George 22.10.2010

Umtamvuna 06.10.2012

Misgund 15.01.2011

Slender lithophyte 70–300mm tall. **Basal leaves** 5–10, green, semi-erect, strap-shaped to narrowly elliptic, up to 90mm long; **cauline leaves** reduced to dry sheaths. **Inflorescence** lax to dense, 5–15-flowered. **Flowers** white to mauve, petals frequently darker than sepals. Median sepal very shallowly hooded, half-moon-shaped, with lateral extensions somewhat reflexed, 6–8 × 8–14mm; spur straight, parallel to ovary, 2–3mm long; lateral sepals projecting away, oblong, 7–8mm long; petals narrowly spear- to strap-shaped from a broad base; lip more or less projecting away, spear- to narrowly egg-shaped, 5–10mm long.

| A | S | O | N | D | J | F | M | A | M | J | J |

Locally common in stony soil, on rocks or in rock crevices, often along streams, or in half-shade, from near sea level to 2,000m. Pollinated by flies. Flowers mainly between September and November, but also in all other months of the year.

Disa salteri

Cape Peninsula 31.03.1993

Oudtshoorn 22.04.1993

Oudtshoorn 14.04.2012

Oudtshoorn 12.04.2012

Oudtshoorn 22.04.1993

Slender, grass-like terrestrial up to 600mm tall. **Radical leaves** developing after flowering, linear, shortly stalked, up to 100mm long; **cauline leaves** completely sheathing, 20mm long. **Inflorescence** lax, 1-sided, 5–20-flowered. **Flowers** brown or bronze. Median sepal hooded, 2.5–5mm tall, 1–2mm deep; spur slender, cylindrical, usually ascending near the tip, 20–30mm long; lateral sepals spear-shaped, 2–4mm long; petals oblong, 2–3mm tall; lip linear, 3–4mm long.

A	S	O	N	D	J	F	M	A	M	J	J

Rare in dry conditions in stony soil derived from sandstone or shale, from near sea level to 1,500m. Flowers between March and April, stimulated by fire.

Disa sanguinea

Sehlabathebe (Lesotho) 07.01.2009

Sani Pass 17.01.2014

Sani Pass 17.01.2014

Sani Pass 17.01.2014

Robust terrestrial 200–800mm tall. **Leaves** cauline, spear-shaped, up to 150mm long. **Inflorescence** dense, many-flowered. **Flowers** with sepals deep red, petals pale green, lip red. Median sepal spherical, 5mm in diameter; spur cylindrical, ascending abruptly from the rear of the galea, 2–3mm long; lateral sepals oblong, 4mm long, semi-erect next to galea; petals included in galea, 5mm long; lip egg-shaped to obscurely 3-lobed, 3mm long.

A S O N D J F M A M J J

Very rare in damp, highland grassland, from 2,400–2,600m. Flowers in January.

Disa sankeyi

Sentinel 03.03.2009

Sentinel 03.03.2008

Witsieshoek 20.02.2010

Sentinel 09.03.2008

Robust terrestrial 150–300mm tall; sterile shoot present. **Leaves** of fertile shoots 6–12, spear-shaped, up to 150 × 20mm. **Inflorescence** dense, cylindrical, 30–80-flowered. **Flowers** facing downward, greenish, white or yellow, fragrant. Sepals 6.5–7.5mm long; spur hanging, slender or somewhat club-shaped, 3–3.5mm long; petals egg-shaped; lip 7 × 1.6mm.

A S O N D J F M A M J J

Very localized in the Drakensberg, in damp to dry grassland in full sun on basalt or sandstone, from 2,400–3,000m. Pollinated by spider-hunting wasps. Hybridizes rarely with *D. fragrans*. Flowers between February and March.

Disa saxicola

Sabie 02.01.2011

Graskop 13.12.2012

Sabie 23.12.2010

Slender, flexuose lithophyte or terrestrial 100–400mm tall. **Leaves** 2 or 3, cauline but clustered near the base, soft, linear to spear-shaped, up to 70–200 × 5–17mm. **Inflorescence** lax, 6–25-flowered, inflorescence stalk and ovaries mottled with purple. **Flowers** white, sometimes suffused with pink, randomly spotted with purple or blue, spur with green apex. Median sepal rounded to apiculate, galea constricted, 4–8 × 3mm, 3mm deep, margins bending inward; spur cylindrical, straight or gently decurved, 4–10mm long; lateral sepals spreading, narrowly oblong to oblong, 4–8mm long; petals narrowly oblong to oblong, 3.5–4.5mm long; lip linear to spatula-shaped, 3–5mm long.

A S O N D J F M A M J J

Distinguishable from *D. oreophila* by the broader, soft leaves and inrolled galea margins.

Occasional in moss on rock ledges and in rock crevices, occasionally on tree trunks, often growing in half-shade, from 1,200–2,200m. Flowers between December and January.

Disa schizodioides

Prince Albert 06.11.2012

Slender terrestrial or lithophyte 50–250mm tall. **Basal leaves** about 6, narrowly to very narrowly elliptic, 10–25mm long. **Inflorescence** usually single-flowered, occasionally up to 3-flowered. **Flowers** white to violet-mauve. Median sepal hooded, 10–12 × 5mm, 8mm deep; spur rudimentary; lateral sepals projecting away, narrowly to broadly egg-shaped, apiculate, 13–20mm long; petals strap-shaped, 4mm long; lip projecting away, 6 × 5mm, margins curved upward.

A S O N D J F M A M J J

Occasional but often in substantial populations in peaty soil or in rock crevices, from 1,000–2,000m. Flowers between November and January, stimulated by fire.

Disa schlechteriana

Syn: *Herschelia schlechteriana*

Herold 12.12.2010

Herold 12.12.2010

Herold 12.12.2010

Slender, reed-like terrestrial up to 600mm tall; stem often with fibrous remains of old leaves at the base. **Radical leaves** about 10, grooved, 300–400 × 2mm; **cauline leaves** about 9, completely sheathing, 30–50mm long. **Inflorescence** lax, 3–12-flowered. **Flowers** cream with mauve veins. Median sepal erect, hooded, 22–25 × 16mm, 8mm deep; spur decurved near the tip, 30–50mm long; lateral sepals projecting away, spear-shaped to narrowly oblong, 20–25mm long; petals with a strap-shaped 14mm-long limb; lip projecting away, flat, narrowly oblong to strap-shaped, 15–20mm long.

A S O N D J F M A M J J

Rare on dry inland slopes in tall proteoid vegetation, from 300–1,000m. Flowers in December.

Disa scullyi

Elliot 25.01.2010

Elliot 22.01.2005

Elliot 25.01.2010

Elliot 29.01.2009

Slender terrestrial up to 500mm tall, with a slender sterile shoot. **Leaves** of sterile shoot 2–4; fertile shoot with 6–15 cauline leaves. **Inflorescence** semi-dense, 10–20-flowered. **Flowers** facing downward at 45°, 20mm in diameter, pink or white, lip tinged with green. Median sepal hooded, 11–13 × 7.5mm, 5mm deep; spur horizontal, cylindrical, slender, 30–40mm long; lateral sepals projecting away, oblong, 11–13 × 6–8mm; petals twisted to face forward, closing the spur entrance; lip half-hanging, strap-shaped to narrowly elliptic, 10 × 2–5mm.

A S O N D J F M A M J J

Localized and very rare in swamps, from 1,500–2,000m. Pollinated by long-proboscid flies. Flowers between December and January.

Disa similis

Port St Johns 21.10.2011

KZN south coast 27.07.2013

KZN south coast 27.07.2013

KZN south coast 29.09.2011

Slender terrestrial 250–500mm tall. **Leaves** 3–8, cauline, spear-shaped, up to 120 × 15mm. **Inflorescence** lax, 10–30-flowered. **Flowers** pale violet to blue to mauve, petals and lip apex yellow. Median sepal erect, almost round, galea 7–9 × 5.5–9mm, 3mm deep, wide open; spur conical, straight, 3mm long; lateral sepals spreading, oblong to egg-shaped, flat with a slight apical keel and apiculus, 7–9 × 4–6.5mm; petals erect, 3.5–4.5 × 1–1.5mm; lip oblong, 3.5–4.5 × 1–2mm.

A S O N D J F M A M J J

Rare in swamps and marshy grassland on the coast, from near sea level to 500m. Flowers usually between September and November, but in some populations as early as July.

Disa spathulata

Slender terrestrial 120–300mm tall. **Radical leaves** 5–20, linear, semi-erect, 50–150 × 2–4mm; **cauline leaves** usually 3, green at flowering time, sheathing, 20–30mm long. **Inflorescence** lax, 1–5-flowered. **Flowers** varying from maroon to pale lime or green and blue. Median sepal almost spatula-shaped, claw horizontal, 1–3mm long, blade usually shallowly hooded, broadly egg-shaped, 9–20mm tall; spur usually club-shaped, straight or strongly curved downward, 1.5–3mm long; lateral sepals egg-shaped, 6–16mm long; petals with apex unequally 2-lobed, limb linear to strap-shaped, 7–12mm long; lip spatula-shaped, limb linear, 5–35 × 1–2mm, blade obscurely to very deeply trilobed.

Disa spathulata subsp. *spathulata*

Syn: *Herschelia spathulata* subsp. *spathulata*

Op-die-Berg 09.10.2007

Citrusdal 14.09.2007

Nieuwoudtville 22.09.2009   Citrusdal 27.08.2007   Citrusdal 14.09.2007   Vanrhynsdorp 25.08.2010

Subsp. *spathulata* is distinguished as follows: lip with midlobe shorter than 14mm. This subspecies is very variable, both within and between populations; the more common form has large pale reddish or greenish flowers and a rarer form has small purplish flowers.

Subsp. *spathulata* is endemic to the Western and Northern Cape provinces of South Africa.

Occasional and usually in damp places but also in well-drained localities, in shale and on sandstone, in full sun, from near sea level to 1,400m. Pollinated by anthophorid bees. Flowers between August and October.

A S O N D J F M A M J J

Disa spathulata subsp. *tripartita*

Syn: *Herschelia spathulata* subsp. *tripartita*

Villiersdorp 10.09.2008

Louterwater 20.09.2011

Louterwater 20.09.2011

Villiersdorp 10.09.2008

Villiersdorp 10.09.2008

Subsp. *tripartita* is distinguished as follows: lip with midlobe 16–22mm long. Subsp. *tripartita* is endemic to the Western Cape province.

Rare and localized in renosterveld and transitional renosterveld-fynbos habitats, from 600–900m. Pollinated by anthophorid bees. Flowers between August and October.

A S O N D J F M A M J J

Disa stachyoides

Sentinel 14.01.2008

Sabie 26.01.2011

Sentinel 14.01.2008

Josefsdal 08.01.2014

Witsieshoek 13.01.2008

Slender terrestrial 100–400mm tall. **Leaves** linear to spear-shaped, 60–100mm long. **Inflorescence** fairly dense, 15–40-flowered. **Flowers** purple, petals and lip with distal two thirds white with a few purple spots. Median sepal hooded, angled forward, margins bending inward, 4–6 × 3–4mm, 1mm deep; spur horizontal, flattened, with a deep groove on the underside and sharply notched at the tip, 2–6mm long; lateral sepals oblong, 4–7mm long, with large fleshy tapering apiculi; petals narrowly oblong, 3–5mm long; lip narrowly spear- to strap-shaped, 4–5mm long.

A S O N D J F M A M J J

Widespread and common in higher-altitude grassland, from 1,000–2,200m. Flowers between December and February.

Disa stricta

Sani Pass 17.01.2014

Sani Pass 17.01.2014

Sani Pass 17.01.2014

Sani Pass 18.01.2014

Slender terrestrial 150–450mm tall. **Leaves** cauline, rigid, linear, up to 200mm long. **Inflorescence** semi-dense, 10–20-flowered. **Flowers** small, sepals pink to purple, lip darker red than sepals, petals white with blue spots apically. Median sepal 4–5mm tall, galea 2mm wide, 2mm deep, front margins extended and often reflexed; spur ascending, straight, 2.5–5mm long; lateral sepals oblong to broadly elliptic, 4–5mm long, shortly apiculate; petals narrowly oblong, 2.5–4mm long; lip projecting away, oblong to elliptic, 3.5–5mm long.

Distinguishable from *D. dracomontana* by the smaller flowers, lack of a clear venation pattern and shorter lateral sepals.

A S O N D J F M A M J J

Occasional but sometimes locally common on grassy slopes or damp floodplains, from 1,800–2,650m. Flowers between November and January.

Disa subtenuicornis

Riversdale 24.11.1992

Riversdale 24.11.1992

Riversdale 24.11.1992

Fairly robust terrestrial 130–250mm tall. **Leaves** cauline, up to 150mm long. **Inflorescence** dense, almost spherical to cylindrical, up to 25-flowered. **Flowers** facing downward, white, sepals and lip with some purple spots, petals with purple borders. Median sepal hooded, 10mm tall, 7mm deep; spur hanging, more or less conical, 3mm long; lateral sepals narrowly egg-shaped to oblong, 10mm long; petals with a small circular basal lobe; lip arrowhead-shaped, 6mm long.

Distinguishable from *D. tenuicornis* by the absence of pouches at the base of the galea.

A S O N D J F M A M J J

Rare and restricted to damp peaty soils on steep south-east-facing slopes, from 1,200–1,300m. Flowers between mid-November and December, after fire.

Disa telipogonis

Wellington 01.12.2008

Wellington 01.12.2008

Porterville 13.10.2010

Slender, flexuose terrestrial 30–60mm tall. **Leaves** linear, as tall as or overtopping the inflorescence, up to 100mm long. **Inflorescence** flat-topped;
bracts of lower flowers leaf-like and overtopping flowers, those of upper flowers
as tall as flowers. **Flowers** bright yellow with brown veins. Median sepal hooded,
4–6mm long, 3–4mm deep, with a 1–2mm-long terminal hair; spur somewhat
hanging, 2–4mm long, with a dorsal groove; lateral sepals 5–6mm long, with
a 1–3mm-long hair; petals with a knee-bend halfway up, 3mm long; lip spear-
shaped, up to 5mm long.

A S O N D J F M A M J J

Occasional in rock crevices, usually in seepages or near streams, from 600–1,750m. Flowers between September
and early December, stimulated by fire.

Disa tenella

Slender terrestrial 30–150mm tall. **Leaves** erect, spiralling, linear, up to 50mm long. **Inflorescence** dense, 5–35-flowered. **Flowers** white, pink or red, often with darker mottling. Median sepal hooded, 3–5 × 4mm, pouch-shaped basally; spur slender, horizontal from the pouch-shaped base of the median sepal, up to 5mm long; lateral sepals oblong to broadly oblong, 4–6.5mm long; petals very broadly egg-shaped, 2–3mm tall; lip narrowly spear-shaped, 5mm long.

Disa tenella subsp. *pusilla*

Op-die-Berg 18.09.2011

Op-die-Berg 01.09.2010

Op-die-Berg 20.09.2008

Op-die-Berg 20.09.2008

Subsp. *pusilla* is distinguished as follows: flowers white; median sepal 3–4mm long; spur straight, up to 2mm long.

Subsp. *pusilla* occurs in upland areas between Ceres and Vanrhynsdorp in the Western Cape province.

Distinguishable from *D. brachyceras* by the straight spur, 2mm long.

Rare in small populations in damp sandy or clay areas, from 250–1,200m. Flowers between September and October, stimulated by fire.

A S O N D J F M A M J J

Disa tenella subsp. *tenella*

Wellington 08.09.2012

Wellington 08.09.2012

Wellington 05.09.2011

Wellington 01.09.2011

Subsp. *tenella* is distinguished as follows: flowers purple or pale pink to reddish-pink; median sepal 3–5mm long; spur slender, decurved, 2.5–5mm long.

Subsp. *tenella* occurs in lowland sand flats between Strand and Hopefield in the Western Cape province.

Rare on the coastal flats but often in large populations in damp to wet areas, from near sea level to 200m. Pollinated by bees. Flowers between August and September.

Disa tenuicornis

Grabouw 24.10.2009

Grabouw 24.10.2009

Grabouw 23.10.2009

Grabouw 25.10.2009

Grabouw 25.10.2009

Robust terrestrial 120–500mm tall. **Leaves** cauline, linear, erect, up to 180mm long. **Inflorescence** dense, many-flowered. **Flowers** facing downward, white, galea with short red lines inside, lateral sepals purplish below, petals and lip with red margin, spur purple-green. Median sepal hooded, 8–10 tall, 4mm deep, with two 1mm-deep pouches basally; spur hanging from between the pouches, slender, 3–4mm long; lateral sepals 8–10mm long; petals parallel to the anther; lip more or less strap-shaped, 6mm long.

Distinguishable from *D. subtenuicornis* by the presence of pouches at the base of the galea.

A S **O** N D J F M A M J J

Very localized but often in large populations on damp peaty soils, from 700–1,200m. Flowers between late September and October, after fire.

Disa tenuifolia

Villiersdorp 08.11.2008

Villiersdorp 17.11.2008

Grabouw 12.12.2009

Hermanus 16.12.2013

Slender terrestrial 50–300mm tall. **Radical leaves** numerous, linear, 10–15mm long; **cauline leaves** spear-shaped, almost completely sheathing. **Inflorescence** lax, 1–10-flowered. **Flowers** bright yellow. Median sepal reflexed, egg-shaped, flat or shallowly concave, 10–18mm long; spur rudimentary; lateral sepals projecting away, spear-shaped, 12–18mm long, apices often reflexed, with long apiculi; petals strap-shaped, 6–8mm long; lip hanging, linear, often curled apically, 6mm long.

A	S	O	N	D	J	F	M	A	M	J	J

Common and often in large populations in damp localities, often in moss, in the zone of the southeaster clouds, from 100–1,800m. Pollinated by leafcutter bees. Hybridizes rarely with *D. filicornis*. Flowers between November and January, after fire.

Disa tenuis

Cape Peninsula 03.06.2011 Cape Peninsula 21.05.2012 Cape Peninsula 04.06.2011

Kleinmond 10.04.2012 Cape Peninsula 03.06.2011

Slender, grass-like terrestrial up to 500mm tall. **Radical leaves** linear, developing after flowering, up to 300mm long; **cauline leaves** lax, dry, up to 50mm long. **Inflorescence** lax to fairly dense, 20–45-flowered. **Flowers** white, greenish-brown, pale purple or rose. Median sepal shallowly hooded, with a 2.5mm-long bristle-like point, galea 3–6mm tall, 2mm deep; spur barely visible, up to 1mm long; lateral sepals projecting away, elliptic-oblong to narrowly egg-shaped, 3–6mm long, apiculi 0.3–2mm long; petals 2mm long, front margin entire or toothed; lip projecting away to hanging, broadly elliptic to narrowly spear-shaped, 2–3mm long, margins entire or toothed.

This species is variable in the size of the flowers, the length of the apiculi and the degree to which the petals, lip and median sepal are notched.

A S O N D J F M A M J J

Rare, singly or in small populations in dry, stony habitats, from near sea level to 1,000m. Flowers mostly between April and June, stimulated by fire.

Disa thodei

Naudesnek 24.01.2010

Naudesnek 08.01.2009

Naudesnek 08.01.2009

Naudesnek 24.01.2010

Slender terrestrial 150–300mm tall, with a slender sterile shoot up to 50mm long. **Leaves** of fertile shoot cauline, 2 or 3. **Inflorescence** lax, 3–8-flowered; bracts usually overtopping the flowers. **Flowers** 20mm in diameter, white to cream, sometimes mottled with pink, spur greenish. Median sepal hooded, 12–17 × 12mm, 4–5mm deep; spur cylindrical, slender, gradually decurved, 20–40mm long; lateral sepals projecting away, egg-shaped to oblong, 10–15 × 5–8mm, apiculi 1mm long; petals erect, 10–15 × 4–6mm; lip narrowly elliptic, apically decurved, 10–15 × 2–4mm.

Distinguishable from *D. crassicornis* by the lateral sepals that are shorter than 20mm.

Occasional along streams or in damp alpine grassland, from 1,800–3,000m. Flowers between December and January.

Disa triloba

Citrusdal 28.11.2013

Citrusdal 28.11.2010

Citrusdal 28.11.2013

Citrusdal 28.11.2010

Slender terrestrial 60–200mm tall. **Leaves** cauline, lower leaves green, strap-shaped or elliptic, up to 40mm long, upper leaves dry, completely sheathing. **Inflorescence** almost flat-topped, 2–8-flowered. **Flowers** mauve with deeper purple markings. Median sepal shallowly hooded, narrowly egg-shaped, often obscurely 3-lobed, 8–10mm long; spur cylindrical, tapering, straight, horizontal to parallel to the ovary, 5mm long; lateral sepals projecting away, narrowly oblong to narrowly egg-shaped, 8–13mm long; petals partially exserted from the galea, very narrowly spear-shaped, 6mm long; lip very narrowly spear-shaped, 8mm long.

A S O N D J F M A M J J

Rare in the drier mountains, usually as widely scattered individuals, in dry to slightly damp places on rocky slopes, from 1,000–1,500m. Flowers between late November and December, stimulated by fire.

Disa tripetaloides

Houwhoek 26.11.2010

Houwhoek 23.11.2010

Houwhoek 21.11.2010

Kleinmond 12.11.2009

Slender terrestrial 100–600mm tall, spreading by stolons. **Basal leaves** up to 10, spreading, narrowly spear-shaped or narrowly elliptic, up to 140mm long. **Inflorescence** lax, up to 150mm long, 7–25-flowered. **Flowers** white to various shades of pink, median sepal and petals spotted with deep red. Median sepal hooded, egg-shaped, 5–12mm long, 3–5mm deep; spur horizontal or somewhat hanging, conical, 2–3mm long; lateral sepals projecting away, broadly egg-shaped, elliptic or oblong, 8–15mm long; petals strap-shaped to narrowly oblong, 4–5mm long; lip projecting away, curved upward apically, linear, 3–4mm long.

A S O N D J F M A M J J

Distinguishable from *D. aurata* by the white flowers, shorter lateral sepals and longer spur, and from *D. caulescens* by the basal leaves and strap-shaped to narrowly oblong petals with deep red spots.

Widespread on banks of perennial streams, from near sea level to 1,000m. Hybridizes rarely with *D. caulescens*. Flowers between November and January in the Western Cape and between May and November elsewhere.

Disa tysonii

Elliot 25.01.2010

Elliot 25.01.2010

Garden Castle 14.12.2008

Garden Castle 23.01.2010

Garden Castle 04.12.2012

Robust terrestrial 200–600mm tall. **Leaves** cauline, narrowly egg-shaped, up to 100mm long. **Inflorescence** dense, many-flowered. **Flowers** cream-yellow with green veins, galea sometimes flushed with purple-pink, lip dark yellow with purple spots. Median sepal with galea 7–9mm tall, 5mm deep; spur cylindrical to more or less club-shaped, 4–6mm long; lateral sepals projecting away, 6–8mm long, with well-developed apiculi; petals strap-shaped, 6mm long; lip projecting away, 4–6mm long.

A	S	O	N	D	J	F	M	A	M	J	J

Rare in damp, rocky grassland, from 1,800–3,000m. Flowers between December and January.

Disa uncinata

George 11.10.2008   George 05.11.2008   George 05.11.2008

Wellington 12.12.2009   Wellington 27.11.2010

Slender, flexuose terrestrial 100–500mm tall. **Leaves** 3–6, narrowly spear-shaped, often curved, up to 250mm long. **Inflorescence** lax, up to 20-flowered. **Flowers** horizontal, white or cream. Median sepal hemispherical, rounded or apiculate, galea 4.5–8mm tall, 5mm deep; spur horizontal and constricted basally, more or less club-shaped apically, 2–4mm long; lateral sepals projecting away, oblong, 5–8mm long; petals curved, 4mm long; lip spear-shaped, 4mm long.

A S O N D J F M A M J J

Common along mountain streams and in wet rock-face habitats that receive water run-off, usually in half-shade, from 100–1,850m. Flowers between October and January, stimulated by fire.

Disa uniflora

Porterville 02.02.2010

Cape Peninsula 19.01.2011

Porterville 04.01.2010

Cape Peninsula 02.02.2006

Slender to fairly robust, flexuose to erect terrestrial 150–600mm tall, spreading by stolons. **Basal leaves** strap- to narrowly spear-shaped, up to 250mm long. **Inflorescence** generally with 1, sometimes up to 3, and very rarely more flowers. **Flowers** with lateral sepals and lip carmine, galea orange or yellow inside with red spots and forked carmine veins, petals pale carmine basally, blade yellow with red spots. Median sepal hooded, 20–60 × 15–20mm, 15mm deep; spur somewhat hanging, parallel to the ovary, conical, 10–15mm long; lateral sepals projecting away, narrowly egg-shaped, flat, 35–65mm long; petals narrowly egg-shaped to rhomboid, 20–25mm long; lip projecting away, reflexed apically, linear, 20–25mm long.

A S O N D J F M A M J J

This species is variable in the size of the plants and in the shape and coloration of the sepals and petals.

Common along perennial streams and in perennially wet seepages over cliffs, from near sea level to 1,450m. Pollinated by the mountain pride butterfly. Hybridizes very rarely with *D. caulescens*. Flowers between January and March.

Disa vaginata

Grabouw 02.12.2009

Cape Peninsula 19.12.2007

Grabouw 18.12.2009

Porterville 29.10.2009

Cape Peninsula 22.12.2009

Slender terrestrial 50–200mm tall. **Leaves** cauline, lower 2–5 clustered basally, strap- to spear-shaped, up to 50mm long, remainder almost entirely sheathing. **Inflorescence** dense, almost flat-topped, 3–10-flowered. **Flowers** rose-pink with darker spots on lateral sepals and petals. Median sepal elliptic, shallowly hooded, 5–6mm long; spur cylindrical, straight, parallel to the ovary, 5–6mm long; lateral sepals projecting away, oblong, 5–6mm long; petals narrowly egg-shaped, 4mm long; lip spear-shaped, 4mm long.

Distinguishable from *D. glandulosa* by the absence of hairs.

A	S	O	N	D	J	F	M	A	M	J	J

Occasional and as isolated plants or small populations in dry stony soil, in moss on rocks or in damp rock crevices, from 300–1,800m. Self-pollinating. Flowers between late October and December, stimulated by fire.

Disa vasselotii

Riversdale 24.11.1992

Riversdale 24.11.1992

Swellendam 16.12.2012

Swellendam 28.12.2012

Slender terrestrial 80–250mm tall. **Basal leaves** 4–10, elliptic, 10–15mm long. **Inflorescence** almost flat-topped, dense, 1–10-flowered. **Flowers** white to pink, sepals often with deeper pink veins on underside, petals with pink and yellow horizontal bars. Median sepal hooded, 9–12 × 8–10mm, 6–8mm deep; spur somewhat hanging, slender, often curved downward, 2–4mm long; lateral sepals projecting away, egg-shaped, 9–15mm long; petals narrowly oblong to narrowly egg-shaped, 6mm long; lip projecting away, strap-shaped, 5mm long.

A	S	O	N	D	J	F	M	A	M	J	J

Rare but sometimes locally abundant in damp, peaty soil on south-eastern slopes, from 600–1,300m. Flowers between November and December, after fire.

Disa venosa

Porterville 08.12.2009

Porterville 08.12.2009

Porterville 08.12.2009

Porterville 08.12.2009

Porterville 19.12.2009

Slender terrestrial 300–500mm tall. **Basal leaves** narrowly spear-shaped, up to 60mm long. **Inflorescence** lax, 2–6-flowered. **Flowers** pale pink to dark pink with darker veins, petals and lip paler, petals with horizontal bars. Median sepal narrowly egg-shaped, margins often reflexed, 16–18mm tall; spur barely visible, conical; lateral sepals projecting away, more or less reflexed apically, narrowly oblong, 12–23mm long; petals narrowly egg-shaped to rhomboid, 8mm long; lip linear, 7mm long.

Distinguishable from *D. racemosa* by the narrow egg-shaped median sepal.

A S O N D J F M A M J J

Widespread but generally quite rare, although occasionally in fairly large populations in swampy areas, from near sea level to 1,700m. Pollinated by carpenter bees. Flowers between November and December, stimulated by fire.

Disa venusta

Bredasdorp 11.12.2010

Bredasdorp 15.12.2013

Bredasdorp 25.12.2012

Bredasdorp 25.12.2012

Bredasdorp 23.12.2012

Slender, reed-like terrestrial 300–600mm tall. **Radical leaves** 6–12, rigid-erect, less than 2mm wide; **cauline leaves** dry, 15–40mm long. **Inflorescence** lax, 2–12-flowered. **Flowers** blue to purple, lip paler or white. Median sepal hooded, 10–18mm tall, 4–8mm deep; spur horizontal, conical, 1.5–3mm long; lateral sepals projecting away, oblong to narrowly oblong, 12–16mm long; petals with a linear limb, 6mm long, apically expanded to form a fan up to 3mm wide or unequally 2-lobed; lip more or less egg-shaped, entire to variably torn, shorter than lateral sepals, decurved, 7–12mm long.

The concept of this species is rather unsatisfactory and further research is needed to clarify relationships and species boundaries.

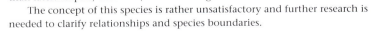

| A | S | O | N | D | J | F | M | A | M | J | J |

Rare and with a disjunct distribution, in damp localities in full sun, from near sea level to 1,200m. Flowers between late November and January.

Disa versicolor

Sabie 01.02.2011

Verloren Vallei 03.01.2011

Verloren Vallei 29.01.2014

KZN south coast 11.10.2013

Moderately robust terrestrial 200–500mm tall, producing a sterile shoot. **Leaves** of sterile shoot usually 5, robust; fertile shoot with 6–15 cauline leaves. **Inflorescence** 17–22mm wide, dense, 50–105-flowered; bracts often dry, 10mm long. **Flowers** with buds pale pink, mottled with pink or maroon, open flowers from yellowish through brown to red, lip and petals green from purple bases, sepals and lip turning brown soon after flowering. Median sepal hooded, 4.5–6.5 × 3.5–4.5mm, 3–4.5mm deep; spur usually sharply deflexed basally, 5–7mm long; lateral sepals projecting away, narrowly oblong to spear-shaped, 4.5–6.5 × 2.5–3.5mm; petals partially included in galea, 4–5 × 3–3.5mm; lip strap-shaped to narrowly spear-shaped, 4.5–5.5 × 1mm.

A S O N D J F M A M J J

Distinguishable from *D. maculomarronina* by the flowers that turn brown shortly after opening.

Widespread and locally common in damp to dry grassland, from near sea level to 2,400m. Pollinated by anthophorid bees. Flowers usually between December and February, but from October onwards at low altitudes.

Disa vigilans

Mokobulaan 02.01.2011

Mokobulaan 16.01.2010

Mokobulaan 02.01.2011

Slender terrestrial 160–280mm tall. **Leaves** 7–9, cauline, rigid, linear to spear-shaped, 50–90 × 3–4mm. **Inflorescence** lax, 7–16-flowered. **Flowers** white suffused with carmine-pink and green, blotched and speckled with maroon. Median sepal narrowly hooded, apiculate, 10–14 × 5–7mm; spur straight, cylindrical, 12–16mm; lateral sepals projecting away, spear-shaped, 10–14 × 3.0–3.5mm; petals narrowly egg-shaped, 5 × 3mm; lip narrowly spear-shaped, slightly decurved, 12–16mm.

| A | S | O | N | D | J | F | M | A | M | J | J |

Very rare in well-drained, exposed, rocky grassland, from 2,000–2,200m. Flowers between December and January.

Disa virginalis

Citrusdal 11.10.2011

Citrusdal 11.10.2011

Citrusdal 11.10.2011

Citrusdal 11.10.2011

Slender, rather flexuose lithophyte 45–150mm tall. **Basal leaves** 3–6, forming a rosette, linear to spear-shaped, up to 40 × 4mm; **cauline leaves** reduced, brown-veined. **Inflorescence** single-flowered. **Flower** white except for conspicuous vertical purple lines on petal blades and a green galea sac. Median sepal hooded, triangular in side view, laterally compressed, 10–15mm tall, 4–6mm deep; spur absent; lateral sepals spreading, egg-shaped, 10–17 × 8–15mm; petal limbs reflexed along base of the galea, 4–6 × 1mm, petal blades facing forward, egg-shaped, 5 × 3mm; lip spreading, strap- to spear-shaped, 10–15 × 4–6mm.

Distinguishable from *D. maculata*, with which it may co-occur, by the white flower colour, shorter, broader lateral sepals and by the narrowed front of the galea.

Very rare on relatively dry rock faces and ledges, always south-facing, from 800–1,000m. Flowers in October, stimulated by fire.

Disa woodii

Hilton 24.10.2011

KZN south coast 30.09.2012

KZN south coast 10.10.2013

KZN south coast 10.10.2013

Slender to robust terrestrial 150–700mm tall, rarely producing a sterile shoot. **Leaves** of fertile shoot 6–25, linear to strap-shaped, up to 250 × 40mm (far fewer and smaller if a sterile shoot is present). **Inflorescence** dense, cylindrical, 40–150-flowered. **Flowers** bright yellow, less than 10mm in diameter. Sepals 5–7mm long; median sepal erect; spur abruptly hanging, usually not or just reaching base of median sepal, 0.8–1.5mm long, rarely up to 3mm; petals curved over anther; lip strap-shaped to linear, often wavy in the middle, 6 × 1mm.

Distinguishable from *D. chrysostachya* and *D. polygonoides* by the yellow flowers and shorter spur that rarely reaches below the base of the median sepal.

| A | S | O | N | D | J | F | M | A | M | J | J |

Common in grassland, roadside embankments and damp conditions, from near sea level to 1,400m. Self-pollinating. Flowers between September and November.

Disa zuluensis

Dullstroom 08.01.2011

Dullstroom 28.01.2012

Dullstroom 28.01.2012

Dullstroom 28.01.2012

Moderately robust terrestrial 500mm–1m tall, producing a sterile shoot. **Leaves** of sterile shoot 155–275 × 15–27mm; leaves of fertile shoot cauline, up to 100mm long. **Inflorescence** fairly dense, 5–15-flowered; bracts slightly longer than the ovaries. **Flowers** facing outwards horizontally, various shades of pink, occasionally lightly spotted, petals white suffused with pink, spur often green apically. Median sepal hooded, 10–15mm tall, 2–3mm deep; spur slender, sharply deflexed, 30–50mm long; lateral sepals projecting away, 10–13mm long, apiculus 1–2mm long; petals 8–13mm long, partially included in galea; lip strap-shaped, somewhat hanging, 13mm long.

Very rare in swampy areas and on stream banks, from 1,000–2,000m. Flowers between December and January.

Disperis anthoceros

Tzaneen 22.02.2014

Lydenburg 14.02.2010

Lydenburg 18.02.2011   Lydenburg 14.02.2010   Tzaneen 22.02.2014

Very slender terrestrial 90–270mm tall. **Leaves** 2, cauline, opposite, egg-shaped, sometimes purple beneath, 25–45 × 20–30mm. **Inflorescence** almost flat-topped, 1–4-flowered, rarely up to 8-flowered. **Flowers** white, petals streaked with green inside. Median sepal forming a conspicuous slender, elongate, tubular spur, 8–11mm long; lateral sepals obliquely egg-shaped to almost round, 5–8 × 3–5mm, spurs conical, 0.75mm long; petals curved, 6–7 × 4–5mm; lip with a short folded limb; appendage 2-lobed at the tip.

Var. *anthoceros* is widespread in sub-Saharan Africa and is the only variety occurring in the region.

A S O N D J **F M** A M J J

Sporadic in leaf litter on forest floor, often on rocky slopes or along streams, also in avocado plantations, from 600–1,800m. Flowers between February and March.

Disperis bodkinii

Cape Peninsula 18.08.2008

Cape Peninsula 11.08.2013

Cape Peninsula 15.08.2007

Cape Peninsula 24.08.2012

Cape Peninsula 24.07.2012

Tiny hairy terrestrial 40–100mm tall. **Leaves** 2, cauline, alternate, egg-shaped to almost circular, margins thinly hairy, 8–25 × 5–17mm, purple-brown below. **Inflorescence** 1- or 2-flowered. **Flowers** with sepals green, petals white. Median sepal with an erect hood 2–3mm long; lateral sepals 2–3mm long, with spurs pouch-shaped, 0.5mm long; petals 2.5 × 1.5mm; lip very narrow; appendage oblong, notched at apex, papillate.

A S O N D J F M A M J J

Very rare in fynbos and plantation forests, from near sea level to 1,300m. Self-pollinating. Flowers between July and October, stimulated by fire.

Disperis bolusiana

Slender, hairy terrestrial 60–150mm tall. **Leaves** 2 or 3, cauline, alternate, egg-shaped, fringed with hairs, sometimes purple below, 6–25 × 3–17mm. **Inflorescence** single-flowered. **Flower** greenish-yellow or white, with a few green spots on petals within. Median sepal with a sac-like hood ascending at 45° and 3–13mm deep; lateral sepals spreading, with a pouch-shaped spur in the middle; petals 4–6 × 4–5mm; lip narrow, from a stalked funnel-like base; appendage triangular, hairy at the apex.

Disperis bolusiana subsp. *bolusiana*

Piketberg 30.08.2010

Villiersdorp 23.08.2010

Villiersdorp 27.08.2007

Villiersdorp 23.08.2010

Villiersdorp 30.08.2007

Subsp. *bolusiana* is distinguished as follows: flowers greenish-yellow; median sepal with hood about as deep as tall, 3–6mm deep; lip claw 5mm long, lip appendage 1.5mm long.

Subsp. *bolusiana* is restricted to the Western Cape south of Citrusdal.

Occasional in renosterveld and fynbos, mainly from 60–700m. Pollinated by oil-collecting bees. Flowers between August and September.

Disperis bolusiana subsp. *macrocorys*

Syn: *Disperis macrocorys*

Wuppertal 24.09.2012

Wuppertal 10.09.2012

Wuppertal 15.09.2012

Wuppertal 15.09.2012

Subsp. *macrocorys* is distinguished as follows: flowers white to pale yellow; median sepal with hood ascending at 45° and upcurved at apex, 1.5–2 times as deep as tall, 9–13mm deep; lip claw 6mm long, lip appendage 3mm long.

Subsp. *macrocorys* is endemic to the Cederberg and adjacent mountains.

Rare, under shrubs in renosterveld, from 500–1,000m. Pollinated by oil-collecting bees. Flowers between September and October.

A S O N D J F M A M J J

Disperis capensis

Slender terrestrial 70–650mm tall. **Leaves** 2, cauline, alternate, spear-shaped, without hairs or with few coarse, stiff hairs, 20–90 × 2.5–6mm. **Inflorescence** single-flowered. **Flower** with sepals green, petals usually magenta with darker margins and papillae but several pale colour forms have been observed. Median sepal hooded, with a 9–20mm long, tail-like appendage at the apex; lateral sepals at first horizontally spreading, then recurved, often twisted backward, with a tail-like appendage up to 20mm long, spurs pouch-shaped, 1.5mm long; petals 8–15 × 6–10mm; lip narrowly spatula-shaped; appendage upcurved, triangular.

Disperis capensis var. *brevicaudata*

Houwhoek 10.09.2012

Houwhoek 10.09.2012

Houwhoek 10.09.2012

Var. *brevicaudata* is distinguished as follows: habit smaller; median sepal boat-shaped, not markedly pouch-shaped; sepals, petals and lip usually without hairs or shortly hairy; sepal tails 2.5–3mm long, shorter than sepals; petals oblong, 4–5mm wide; lip without an appendage.

Var. *brevicaudata* occurs sporadically in the Cape Floral Region.

Sporadic and sometimes together with var. *capensis*, from 350–1,100m. Flowers between September and December.

A S O N D J F M A M J J

Disperis capensis var. *capensis*

Piketberg 09.09.2008

Cape Peninsula 26.07.2012

Cape Peninsula 15.08.2007

Var. *capensis* is distinguished as follows: habit taller; median sepal forming an ascending hood; sepals, petals and lip shortly hairy; sepal tails 9–20mm long, longer than sepals; petals broadly oblong, 6–10mm wide; lip with an appendage.

Var. *capensis* is widespread in the Cape Floral Region.

Common on damp sandstone, often in seepages, from 10–1,100m. Pollinated by carpenter bees. Flowers between July and September.

A S O N D J F M A M J J

Disperis cardiophora

Witsieshoek 07.02.2007

Witsieshoek 21.02.2010

Witsieshoek 29.01.2008

Witsieshoek 29.01.2008

Slender terrestrial 110–260mm tall. **Leaf** 1, basal, almost circular, margins translucent, 10–17 × 8–19mm. **Inflorescence** 1-sided, 6–20-flowered. **Flowers** greenish-white to pink, tips of sepals and petals magenta, petals spotted with pink within. Median sepal with a slightly decurved sac-like spur 1–1.5mm long; lateral sepals spreading, 4–5 × 2.5–3mm, spurs 1–1.5mm long; petals 4 × 2–3mm; lip reflexed over the column, with a short egg-shaped limb; appendage erect, oblong, longer than the limb.

Distinguishable from *D. renibractea* by the narrower flowers (12–15mm) and orientation of the petals that obscure the inside when the flower is viewed from the front.

| A | S | O | N | D | J | F | M | A | M | J | J |

Fairly common in montane to subalpine grassland, on slopes or in seepage areas and along watercourses, from 1,000–2,500m. Pollinated by oil-collecting bees. Flowers between December and February.

Disperis circumflexa

Slender terrestrial 80–220mm tall. **Leaves** 2, cauline, alternate, linear to spear-shaped, 25–90 × 1.5–6mm. **Inflorescence** 1-sided, 2–9-flowered. **Flowers** greenish-yellow or white and green, with darker green spots on petals within. Median sepal with a hood ascending at 45°, 1.5–8mm deep; lateral sepals horizontal, recurved at the tips, egg-shaped, spurs 1.5–2mm long; petals 4–6 × 2–3mm; lip linear, hairy at apex.

Disperis circumflexa subsp. aemula

Syn: *Disperis circumflexa* var. *aemula*

Nieuwoudtville 06.09.2008

Citrusdal 26.08.2007

Nieuwoudtville 07.09.2006

Nieuwoudtville 07.09.2006

Subsp. *aemula* is distinguished as follows: flowers larger, green and white; median sepal with hood deeper (6.5–8mm) than tall (4–6mm).

Subsp. *aemula* occurs in the north-western part of the Cape Floral Region.

Occasional on clay or granite in mountainous areas, from 500–1,000m. Pollinated by oil-collecting bees. Flowers between August and early October.

A S O N D J F M A M J J

Disperis circumflexa subsp. *circumflexa*

Ceres 24.09.2010

Citrusdal 06.10.2007

Citrusdal 16.09.2014

Subsp. *circumflexa* is distinguished as follows: flowers greenish-yellow; median sepal with hood taller (4–6mm) than deep (1.5mm).

  Subsp. *circumflexa* is widespread in the western part of the Cape Floral Region.

Occasional on sandy, clay or granitic soils in low fynbos and renosterveld, from 100–600m. Flowers between August and early October.

A S O N D J F M A M J J

Disperis concinna

Sabie 16.01.2010

Sabie 19.01.2010

Maclear 13.02.2008

Maclear 27.01.2009

Garden Castle 22.01.2008

Very slender terrestrial 150–350mm tall. **Leaves** 3, cauline, alternate, narrowly egg-shaped, 15–32 × 4–9mm. **Inflorescence** 2–5-flowered. **Flowers** 10mm in diameter, pale magenta or white, petals spotted with green within. Median sepal with a sac-like hood 3–4mm deep; lateral sepals spreading, 6–7 × 2.5–3mm, spurs 2–2.5mm long; petals 3–5 × 1.5–2mm; lip with a folded, egg-shaped to oblong limb; appendage broadly oblong, keeled, longer than the limb, the tip bent down; rostellum arms projecting, scarcely twisted at the tips, 3–4mm long.

Distinguishable from *D. wealei* by the projecting rostellum arms, and reliably from *D. cooperi* and *D. tysonii* only by the tip of the lip appendage, which is bent down.

| A | S | O | N | D | J | F | M | A | M | J | J |

Rare in damp grassland, along watercourses or in seepages or rock sheets, from 1,400–2,300m. Flowers between January and February.

Disperis cooperi

Sabie 10.03.2012

Sabie 09.03.2012

Sabie 09.03.2012

Sabie 11.03.2012

Slender terrestrial 140–380mm tall. **Leaves** 3 or 4, cauline, alternate, egg- to spear-shaped, 11–20 × 5–7mm. **Inflorescence** 1-sided, usually 7–11-flowered. **Flowers** white to cerise-pink, lateral sepals whitish or paler pink on the inside, petals cream spotted with green within. Median sepal with a sac-like hood 6mm deep; lateral sepals hairy at the base, 11 × 4mm, spurs 2mm long; petals 5.5 × 3mm; lip with an egg-shaped limb and pointed appendage about twice as long as the limb; rostellum arms 4mm long, bulbous at the base, shortly projecting.

Reliably distinguishable from *D. concinna* and *D. tysonii* only by the lip appendage, which is straight or slightly curved upward and about twice as long as the limb.

| A | S | O | N | D | J | F | M | A | M | J | J |

Very rare and scattered on damp grassy slopes, from 1,200–2,200m. Flowers between February and March.

Disperis cucullata

Darling 29.08.2010

Darling 26.08.2012

Darling 29.08.2010

Darling 25.07.2012

Slender and shortly hairy terrestrial 85–160mm tall. **Leaves** 2, cauline, alternate, elliptic to egg-shaped, margins shortly hairy, 7–25 × 3.5–11mm. **Inflorescence** single-flowered. **Flower** green, petals marked with darker green spots. Median sepal with an erect hood; lateral sepals 8–12mm long, spurs pouch-shaped, strongly upturned apically, 2–3mm long; petals 4–6 × 1–1.5mm; lip narrow from a broad, stalked base; appendage linear, hairy in the distal half.

| A | S | O | N | D | J | F | M | A | M | J | J |

Uncommon in seasonally damp, often sandy places, from 30–600m. Flowers between August and September, stimulated by fire.

Disperis disiformis

Nature's Valley 22.07.2012

Karino 11.07.2012

Nature's Valley 18.07.2013

Karino 12.07.2012

Very slender terrestrial 70–150mm tall. **Leaves** 2, cauline, alternate, egg-shaped to broadly egg-shaped, 10–26 × 4–16mm. **Inflorescence** 1–4-flowered. **Flowers** off-white, suffused with pink or lilac. Median sepal hooded, with a deflexed spur 2–5mm long; lateral sepals spreading, 5 × 2mm, spurs 1mm long; petals 3.5 × 2.5mm; lip with a broadly triangular limb; appendage strap-shaped, apical part with papillae.

Distinguishable from *D. micrantha* by the deflexed, conical spur of 2–5mm long on the median sepal.

A S O N D J F M A M J J

Occasional in shaded spots under trees in relatively dry and open *Acacia* bushveld, in grassland and wet road cuttings, from near sea level to 1,700m. Flowers between May and October.

Disperis fanniniae

Graskop 02.02.2011

Royal Natal National Park 18.01.2008

Ngeli 31.01.2009

Giant's Castle 05.02.2007

Ngeli 29.01.2009

Slender terrestrial 150–400mm tall. **Leaves** 3, cauline, alternate, egg- to spear-shaped, 20–80 × 8–33mm. **Inflorescence** 1–10-flowered. **Flowers** white, flushed pink or green, petals green along margins and spotted with mauve within. Median sepal with a prominent hood 10–16mm deep; lateral sepals spreading, decurved, spurs 1mm long; petals with a subapical lobe, 7–8 × 5–6mm; lip linear, dilated in the centre, lobed and papillose at apex; appendage egg-shaped, shorter than limb.

| A | S | O | N | D | J | F | M | A | M | J | J |

Common among leaf litter in forest and in humus on rocks, also found in pine, cypress or wattle plantations, from 700–2,100m. Pollinated by oil-collecting bees. Flowers between January and March.

Disperis johnstonii

Gorongosa National Park (Mozambique) 17.04.2013

Gorongosa National Park (Mozambique) 17.04.2013

Tiny terrestrial 45–150mm tall. **Leaves** 2, cauline, alternate, egg-shaped, dark green with white venation, purple beneath, 12–30 × 7–22mm. **Inflorescence** almost flat-topped, 2–5-flowered. **Flowers** with lateral sepals white, petals yellow or pale magenta. Median sepal flat but bending inward; lateral sepals fused in basal third, unevenly rounded, spurs conical, 0.5–1mm long; petals 9–10 × 2–3mm; lip claw linear; blade stalked, terminating in a rounded egg-shaped tip with a papillate crest; appendage 2-lobed, each lobe papillate.

A S O N D J F M A M J J

Very rare in coastal grassland and grassy places under trees, often on rocky slopes, in rich humus or sandy soil, from near sea level to 600m. Flowers between June and July.

Disperis lindleyana

Ngeli 19.12.2010

Ngeli 19.12.2010

Ngeli 20.12.2010

Ngeli 20.12.2010

Very slender terrestrial 140–300mm tall. **Leaf** 1, cauline, halfway up the stem, egg-shaped, heart-shaped at base, glossy, sometimes purple beneath, 20–55 × 13–40mm. **Inflorescence** 1–4-flowered. **Flowers** white suffused with pale green, petals with pink spots within. Median sepal 9 × 5mm, with a slightly decurved sac-like hood 3–4mm deep; lateral sepals at first ascending, then sharply deflexed, 10–12 × 4–6mm, spurs 2–2.5mm long; petals 6–7 × 3–4mm; lip with a broad linear limb; appendage ascending, oblong, apex hairy, shorter than the limb.

A S O N D J F M A M J J

Common in leaf litter in forested habitats and in well-drained, often rocky places, sometimes in humus on rocks, as well as in pine and wattle plantations, from 500–2,100m. Flowers mostly between December and January.

Disperis macowanii

Nature's Valley 03.05.2013

Cape Peninsula 10.06.2012

Elliot 25.02.2010

Nature's Valley 03.05.2013

Cape Peninsula 10.06.2012

Elliot 26.02.2010

Tiny, hairy terrestrial 70–120mm tall. **Leaves** 2, cauline, alternate, egg-shaped, fringed with hairs, purple beneath, 15–20 × 10–16mm. **Inflorescence** single-flowered. **Flower** white or mauve, petals marked with magenta spots, lip apex magenta. Median sepal with a tall conical hood; lateral sepals spreading, spurs pouch-shaped, conical, 1.5mm deep; petals 4 × 1.5mm; lip oblong, short; appendage linear, apex 3- or 4-lobed, several times longer than limb.

A	S	O	N	D	J	F	M	A	M	J	J

Occasional under vegetation in damp and partially shaded locations or more commonly on rocky slopes or cliffs, in organically rich soils, often among mosses or lichens, from near sea level to 2,500m. Self-pollinating. Flowers between December and March in the summer-rainfall region and between May and August in the winter-rainfall region.

Disperis micrantha

Lydenburg 14.02.2010

Lydenburg 22.02.2011

Graskop 08.02.2013

George 19.03.2013

Very slender terrestrial 80–180mm tall. **Leaves** 2 or 3, cauline, alternate, broadly egg-shaped, heart-shaped at base, 25–55 × 14–36mm. **Inflorescence** 2–7-flowered. **Flowers** very small, off-white, suffused with pale green and/or pink. Median sepal hooded, with a blunt, suberect spur 1mm long; lateral sepals not spreading widely, 1.5mm long, spurs 0.5mm long; petals 2.5 × 2mm; lip with a narrow base and two divergent lobes on the limb; appendage broadly triangular, apex with long papillae.

Distinguishable from *D. disiformis* by the blunt, suberect spur of 1mm long on the median sepal.

A S O N D J F M A M J J

Occasional in leaf mould in moist, shaded spots under trees, often among rocks, sometimes in humus on rocks or tree trunks, from 100–1,800m. Flowers between February and April.

Disperis oxyglossa

Sentinel 03.03.2009

Sentinel 28.02.2007

Sentinel 28.02.2007

Cathedral Peak 28.02.2008

Slender terrestrial 160–330mm tall. **Leaves** 2–4, cauline, alternate, egg-shaped, 10–35 × 4–8mm. **Inflorescence** 2–7-flowered. **Flowers** pink, lateral sepals paler on the inside and with a greenish centre, petals cream with green spots. Median sepal forming a shallow hood 4–5mm deep; lateral sepals horizontally spreading, 4–17 × 4.5–7mm, spurs 3mm long; petals 8–10 × 2.5–4mm; lip spear-shaped to tapering to a long tip, reflexed over the column; appendage curved, linear, with a pair of broad, triangular wings at the base; rostellum arms 3.5mm long, projecting, flexed upward in outer third.

A S O N D **J F M** A M J J

Occasional in damp grassy places at the base of slopes and in seepages or along watercourses, from 1,200–2,700m. Flowers between January and early March.

Disperis paludosa

Grabouw 07.12.2009

Cape Peninsula 27.10.2007

Cape Peninsula 28.10.2007

Cape Peninsula 30.10.2007

Slender terrestrial 200–460mm tall. **Leaves** usually 3, cauline, alternate, linear to spear-shaped, 15–55 × 1.5–7mm. **Inflorescence** 2–4-flowered. **Flowers** magenta and green, petals spotted with purple, rarely white. Median sepal hooded, 2–3mm deep, hairy within; lateral sepals at first spreading horizontally, then sharply deflexed in distal half, spurs 1.5–3mm long; petals 10–11 × 3mm; lip with a papillose knob.

A S O N D J F M A M J J

Occasional in marshy ground on sandstone, from 20–1,300m. Flowers between late October and December, after fire.

Disperis purpurata

Slender and hairy terrestrial 90–140mm tall. **Leaves** 2, cauline, alternate, egg-shaped, fringed with hairs, purple below, 19–30 × 9–25mm. **Inflorescence** single-flowered. **Flower** magenta, sometimes pure white or white flushed with pink, petals with green spots, ovary purple. Median sepal hooded, 7–8mm deep; lateral sepals 10–15mm long, spurs pouch-shaped, 2–6mm deep; petals 4–8 × 4–5mm; lip narrowly funnel-shaped; appendage triangular, apex hairy.

Disperis purpurata subsp. pallescens

Eksteenfontein 08.08.2013

Eksteenfontein 24.08.2013

Eksteenfontein 24.08.2013

Eksteenfontein 24.08.2013

Subsp. *pallescens* is distinguished as follows: flowers pale pink; lateral sepals paler pink to nearly white, spurs 5–6mm long; petals bent like a knee; rostellum arms 4.5–5mm long.

Subsp. *pallescens* is endemic to the Richtersveld in the Northern Cape.

Known from only a few populations in schist soil under shrubs, from 850–1,350m. Flowers between August and early September.

Disperis purpurata subsp. *purpurata*

Matjiesfontein 19.09.2011

Matjiesfontein 19.09.2013

Matjiesfontein 18.09.2013

Nieuwoudtville 17.09.2011

Subsp. *purpurata* is distinguished as follows: flowers magenta or pure white; lateral sepals with spurs 2–3mm long; petals curved; rostellum arms 2.5mm long.

Subsp. *purpurata* occurs in the Western and Northern Cape, with one record from Humansdorp in the Eastern Cape.

Occasional but often in large colonies in drier, mountainous inland areas, in shaded, moist spots at the base of rocks or among shrubs on rocky slopes, usually in seepage areas, from 600–1,700m. Pollinated by oil-collecting bees. Flowers between August and October.

A S O N D J F M A M J J

Disperis renibractea

Sabie 24.01.2014

Sabie 06.02.2013

Witsieshoek 16.01.2008

Sentinel 26.01.2008

Sabie 26.01.2011

Slender terrestrial 90–280mm tall. **Leaf** 1, basal, egg-shaped, margins translucent, 19–40 × 14–26mm. **Inflorescence** 1-sided, 4–17-flowered. **Flowers** greenish, pinkish or brownish, petals green and white with maroon spots within, lip maroon basally. Median sepal forming a shallow decurved hood 1–2mm long; lateral sepals spreading, 6.5–8 × 3.5–4mm, spurs 1.5–2mm long; petals 7 × 3.5mm; lip with a linear to strap-shaped limb that is winged on both margins; appendage triangular, apex hairy.

Distinguishable from *D. cardiophora* by the broader flowers (23–27mm) and the orientation of the petals, which do not obscure the inside of the flower when viewed from the front.

A S O N D J F M A M J J

Sporadic but sometimes locally abundant in moist montane grassland on slopes, from 1,300–2,800m. Pollinated by oil-collecting bees. Flowers between December and February.

Disperis **287**

Disperis stenoplectron

Sentinel 19.02.2010

Witsieshoek 01.02.2007

Sentinel 10.02.2009

Garden Castle 23.01.2010

Garden Castle 09.02.2008

Sentinel 07.02.2007

Slender to fairly stout terrestrial 140–230mm tall. **Leaves** 3 or 4, cauline, alternate, narrowly egg- to spear-shaped, 20–35 × 8–12mm; bracts prominent. **Inflorescence** 1-sided, 3–8-flowered. **Flowers** from dull white to deep purple-pink, sepals with cream at the base, petals cream, spotted with green within. Median sepal with a hood 9mm deep; lateral sepals spreading, 10–12 × 4–5mm, spurs up to 2mm long; petals with a raised ridge at the joint with the median sepal, 5–6 × 3–4mm; lip egg-shaped to oblong with margins bending inward; appendage erect, linear, longer than limb; rostellum arms 3.5mm long, projecting, twisted at the tips.

A S O N D J F M A M J J

Occasional on montane to subalpine moist grassy slopes, from 1,200–2,850m. Pollinated by oil-collecting bees. Hybridizes with *D. tysonii*. Flowers between January and April.

Disperis thorncroftii

Giant's Castle 07.12.2011

Giant's Castle 05.12.2013

Giant's Castle 05.12.2013

Giant's Castle 07.12.2011

Very slender terrestrial 140–260mm tall. **Leaves** 2, cauline, alternate, the lower much larger, egg-shaped, purple beneath, 15–45 × 6–33mm. **Inflorescence** 1–3-flowered. **Flowers** white or lilac, petals with green spots within. Median sepal with a broad shallow hood; lateral sepals spreading, 6–8 × 3–4.5mm, spurs curved, 2–3mm long; petals with a finely toothed outer margin; lip with an egg-shaped limb; appendage short, linear, nearly as long as the limb.

A S **O** **N** **D** J F M A M J J

Very rare in leaf litter in coastal to montane forest, from 250–2,000m. Flowers between November and December.

Disperis tysonii

Garden Castle 24.01.2008 — Garden Castle 28.02.2009 — Garden Castle 07.03.2009

Witsieshoek 20.02.2010 — Witsieshoek 20.02.2010

Slender terrestrial 200–400mm tall. **Leaves** 4 or 5, cauline, alternate, egg- to spear-shaped, 15–30 × 5–8mm. **Inflorescence** 1-sided, 5–20-flowered. **Flowers** pink to bright magenta, sometimes white, lateral sepals paler on inside, petals magenta, spotted with green within, lip green. Median sepal with a tall, sac-like hood 4–5mm deep and often tipped backward; lateral sepals spreading, 6–8 × 2.5–3mm, spurs up to 0.5mm deep; petals 4–6 × 2–2.5mm; lip with an egg-shaped pointed limb; appendage straight, long and narrow, three times the length of the limb; rostellum arms short, 0.75mm long, bulbous at the base, incurled.

Reliably distinguishable from *D. concinna* and *D. cooperi* only by the straight, long, narrow lip appendage.

A S O N D J F M A M J J

Restricted but locally common on damp grassy slopes, often on sandstone or quartzite, from 1,200–2,300m. Pollinated by oil-collecting bees. Flowers between January and March.

Disperis villosa

Stanford 12.09.2006

Ceres 11.10.2007

Wolseley 19.09.2007

Stanford 03.09.2007

Slender and hairy terrestrial 60–180mm tall. **Leaves** 2, cauline, alternate, elliptic to almost circular, fringed with hairs, 6–25 × 3–17mm. **Inflorescence** 1–3-flowered. **Flowers** greenish-yellow. Median sepal with a deep sac-like hood projecting horizontally, 5–7mm deep, hairy; lateral sepals horizontally spreading, then bent sharply inward, 6–7 × 3–4mm, spurs pouch-shaped, arising from upper half of lateral sepals, 2.5mm deep; petals 3 × 1.5mm; lip long and narrow; appendage ascending, narrowly boat-shaped, apex hairy.

A S O N D J F M A M J J

Common on clay or granite soils, from 10–760m. Pollinated by oil-collecting bees. Flowers between August and early October, stimulated by fire.

Disperis virginalis

Barberton 14.02.2011

Barberton 15.02.2011

Warburton 21.03.2010

Graskop 07.02.2013

Very slender terrestrial 85–250mm tall. **Leaves** 2, cauline, opposite, broadly egg-shaped, purple beneath, 20–70 × 15–25mm. **Inflorescence** 1–3-flowered. **Flowers** mauve, petals streaked with maroon, lip appendage yellow. Median sepal narrow, 10–13mm long, with a slightly compressed rounded but shallow hood 2mm deep; lateral sepals fused at the base, then spreading, 8–10 × 6–8mm, spurs conical, 1.5mm long; petals wide with two broad lobes on outer margin; lip with limb small, linear and appendage long and terminating in 4 spreading lobes.

A S O N D J F M A M J J

Rare in leaf mould in forest but common in pine plantations, from 1,600–2,000m. Flowers between February and April.

Disperis wealei

Sentinel 09.02.2009

Sentinel 28.02.2012

Sentinel 28.02.2012

Sentinel 10.02.2009

Garden Castle 27.01.2008

Slender terrestrial 140–270mm tall. **Leaves** 2–4, cauline, alternate, egg-shaped, 11–20 × 5–7mm. **Inflorescence** 1–4-flowered. **Flowers** 20–25mm in diameter, white, petal edges with a band of green spots. Median sepal with a bell-shaped hood 4–5mm deep; lateral sepals spreading, ending in a fine point, 9–11 × 4–5mm, spurs 1.5–2.5mm long; petals 6–8 × 4–6mm; lip egg-shaped, hooded at the tip; appendage narrower and pointed, about as long as the limb; rostellum arms sharply deflexed and bent like a knee, 5mm long.

A small form of this species, 50–100mm tall and with flowers 10–12mm in diameter, has often been confused with *D. concinna* but can be distinguished by the deflexed rostellum arms, which are bent like a knee.

A S O N D J F M A M J J

Fairly common in wet grassy situations, seeps or flushes or along watercourses, sometimes at forest margins, from 1,500–2,700m. Pollinated by oil-collecting bees. Flowers between January and February.

Disperis woodii

Assagay 27.03.2005

Howick 04.05.2014

Assagay 27.03.2005

Slender terrestrial 60–150mm tall. **Leaves** 2, almost near the base, subopposite, egg-shaped, dark green with silvery veins, 10–20 × 6–12mm. **Inflorescence** single-flowered. **Flower** whitish or tinged with pink, median sepal darker or olive, lateral sepals marked with longitudinal pink streak. Median sepal with an erect, tubular spur up to 8–11mm long; lateral sepals tapering to a fine tip, spurs shallow, 1mm long; petals 4.5 × 3mm; lip with a long basal claw and egg-shaped limb; appendage extending into the spur, three times as long as the limb, bifid at apex.

A S O N D J F M A M J J

Rare in damp, usually sandy grassland, sometimes within tussocks, from sea level to 1,800m. Flowers between March and August.

Dracomonticola virginea

Naudesnek 08.12.2013

Naudesnek 13.12.2008

Naudesnek 20.11.2011

Sentinel 02.01.2009

Slender terrestrial up to 250mm tall. **Leaves** 2, one basal, elliptic, up to 40 × 22mm, and a much smaller one halfway up the stem, narrowly triangular, up to 20 × 2mm. **Inflorescence** short and dense, 1–8-flowered, up to 50mm long. **Flowers** with sepals white, petals pale pink with dark pink margins. Sepals similar, 8–15 × 3–6mm; petals up to 4 × 2mm; lip spatula-shaped to fiddle-shaped, 4–6 × 3–4mm.

| A | S | O | N | D | J | F | M | A | M | J | J |

Localized on grassy ledges and in rock crevices, usually over basalt, from 1,500–3,100m. Flowers between October and January.

Eulophia angolensis

White River 13.01.2010

Alkmaar 13.01.2010

KZN south coast 08.12.2012

Robust terrestrial 500mm–1.5m tall. **Leaves** pleated, stiffly erect, fully developed at flowering time, up to 900 × 50mm. **Inflorescence** lax, 10–40-flowered. **Flowers** spreading, bright lemon-yellow, variously tinged with olive and purplish-brown. Sepals similar, markedly erect, 19–24mm long; petals curving over the lip, slightly shorter than sepals; lip 3-lobed, with a crest of 3–5 blades, usually low basally and rising to more than 1.5mm tall on midlobe; spur shortly conical to almost absent; column 9–14mm long, column foot 4–10mm long.

| A | S | O | N | D | J | F | M | A | M | J | J |

Locally common in swampy soil, often in open grassy places, from near sea level to 1,400m. Pollinated by carpenter bees. Flowers between December and March.

Eulophia calanthoides

Luneburg 22.01.2014

Luneburg 14.01.2013

Giant's Castle 07.02.2009

Luneburg 14.01.2013

Moderately robust to robust terrestrial 300–750mm tall. **Leaves** thin-textured, fully developed at flowering time, up to 700 × 60mm. **Inflorescence** lax, 7–25-flowered. **Flowers** more or less nodding, sepals purplish-brown outside, brownish-green within, petals very pale yellow to milky white, tinged with purple near the base, veins marked with numerous minute blue speckles on inner surface, lip very pale yellow with minute dark blue speckles along the main lateral veins, crests bright yellow. Sepals and petals similar, 26–30mm long; sepals spreading; petals covering the column; lip obscurely 3-lobed, with a broadly concave sac 3mm deep beneath the column apex, crests consisting of low ridges in basal third to quarter; spur 4–5mm long; column 4–5mm long, column foot almost absent to absent.

A	S	O	N	D	J	F	M	A	M	J	J

Occasional in bushy and tall herbaceous vegetation on forest margins or in *Leucosidea* thicket, in colonies or as scattered individuals, from 1,200–2,000m. Flowers between December and February.

Eulophia callichroma

Songimvelo 09.10.2012

Songimvelo 09.10.2011

Songimvelo 22.11.2012

Songimvelo 09.10.2011

Fairly robust terrestrial up to 600mm tall; above-ground pseudobulbs apparently absent in South African plants, but recorded elsewhere as well developed, up to 100mm long. **Leaves** absent at flowering time, up to 170 × 10mm. **Inflorescence** lax, branched, 30–50-flowered, arising from an upper node of the pseudobulb. **Flowers** resupinate or not, sepals purplish-brown, petals paler. Sepals deeply concave, 8 × 1.5–2mm; petals similar but slightly smaller, with a distinct keel; lip 3-lobed, with a crest of 5 ridges; spur 4–5mm long; column 2.5–4.5mm long, column foot 1mm long.

A S O N D J F M A M J J

Rare in open sour bushveld, from 100–1,100m. Flowers between September and October.

Eulophia clitellifera

KZN south coast 18.07.2012

KZN south coast 21.07.2012

KZN south coast 27.06.2013

KZN south coast 05.07.2013

Slender terrestrial 200–350mm tall. **Leaves** thick-textured, absent or just emerging at flowering time, up to 240 × 13mm. **Inflorescence** lax, 5–25-flowered. **Flowers** with sepals dull green, tinged with brown and purple, petals and lip dull white with reddish-purple lines on inner surfaces, lip crests bright yellow. Sepals and petals spreading; sepals similar, elliptic, 5–8mm; petals egg-shaped, 6 × 7mm; lip 3-lobed, crests consisting of warty ridges, mainly on convex portion of midlobe; spur 2–5mm long; column 2–4mm long, column foot a little shorter.

A S O N D J F M A M J J

Widespread but always scattered and uncommon in exposed conditions in a variety of different vegetation types: arid succulent scrub, coastal thornveld and grassland, tall and sourveld grassland, and bushveld, from near sea level to 1,700m. Flowers between the end of June and October.

Eulophia coddii

Heidelberg (Gauteng) 18.12.2013

Heidelberg (Gauteng) 20.12.2011

Heidelberg (Gauteng) 20.12.2011

Heidelberg (Gauteng) 12.01.2013

Heidelberg (Gauteng) 20.12.2011

Slender to moderately robust terrestrial 280–420mm tall. **Leaves** partly to fully developed at flowering time, narrowly linear, up to 250 × 5mm. **Inflorescence** lax, 6–15-flowered. **Flowers** with sepals green, heavily suffused with red-brown on inner surface, petals cream with blue veins on inner surface, anther cap rose. Sepals spreading, equal, 12–15mm long; petals overarching the column; lip 3-lobed, crests of 7–9 low warty ridges extending to near midlobe apex; lip midlobe irregularly and finely toothed; spur 2mm long; column 8mm long, column foot 4–5mm long.

A	S	O	N	D	J	F	M	A	M	J	J

Very rare in grassy places and mixed bush, on steep hillsides on sandstone-derived soils, from 1,500–1,900m. Flowers in December.

Eulophia coeloglossa

KZN south coast 09.11.2011

KZN south coast 09.11.2011

KZN south coast 01.12.2013

KZN south coast 08.11.2011

Robust terrestrial up to 600mm tall. **Leaves** stiff, almost fully developed at flowering time, 100 × 10mm. **Inflorescence** lax, 5–10-flowered. **Flowers** with sepals dark pink or maroon, petals pale green tinged with pink, lip yellow tinged with purple near apex and on veins of side lobes. Sepals spreading, equal, the margins rolled inwards to form a cylinder, 14–16mm long; petals broader, 12–16mm long, set close to the column; lip very indistinctly 3-lobed, crests consisting of 2 broad, tooth-like calli in basal half of lip, on either side of a low central blade; spur a rounded, indistinct sac; column 7–10mm long, column foot 4mm long.

A S O N D J F M A M J J

Rather rare in marshes, vleis and moist grassland, near sea level. Flowers between August and December.

Eulophia cooperi

Breyten 25.10.2012

Carolina 03.11.2011

Breyten 26.10.2012

Breyten 25.10.2012

Fairly robust terrestrial 160–370mm tall. **Leaves** rather leathery, scarcely to partly developed at flowering time, reaching 300 × 16mm. **Inflorescence** lax to fairly dense, 6–18-flowered. **Flowers** sometimes non-resupinate at flowering time, sepals pale straw tinged with green, petals straw, lip straw tinged with purple, crests bright yellow. Sepals 19–26mm long; petals as long as or a little shorter than sepals; lip 3-lobed, crests consisting of 3–7 low wavy ridges in basal third and a few stout papillae basally; spur 1.2–2.2mm long; column 7–8mm long, column foot 2–5mm long.

A S **O** N D J F M A M J J

Rare in grassland from 1,200–1,900m. Flowers between October and November.

Eulophia cucullata

Nyoni 23.10.2011

Nyoni 23.10.2011

Nyoni 25.10.2012

Nyoni 22.10.2012

Rather slender terrestrial 250–500mm tall. **Leaves** pleated, absent to partly developed at flowering time, up to 300 × 10mm. **Inflorescence** lax, 3–15-flowered. **Flowers** with sepals brownish-maroon, petals pale purplish-pink, lip white tinged with purplish-pink, crests yellow, sometimes tinged with purple, sac inside yellow with orange and purple spots. Sepals reflexed, 14–16mm long; petals slightly spreading, almost circular, as long as sepals; lip 3-lobed, with a shallow, broadly rounded sac, crests consisting of 2 erect blades near the edge of the sac; column 9–18mm long.

A S O N D J F M A M J J

Frequent to locally abundant in a narrow belt along the coast, in grassy places in bushveld, from near sea level to 400m. Pollinated by carpenter bees. Flowers mainly between October and November.

Eulophia fridericii

Tzaneen 19.12.2013

Tzaneen 21.12.2013

Tzaneen 19.12.2013

Tzaneen 19.12.2013

Slender terrestrial 300–600mm tall. **Leaves** thin-textured, usually fully developed at flowering time, up to 250 × 65mm. **Inflorescence** lax, elongate, 3–20-flowered. **Flowers** with sepals dull lime-green tinged with dark reddish-brown, petals bright lemon-yellow outside with brownish-red lines on pale yellow inside, lip yellow with reddish-brown lines. Sepals and petals spreading; sepals 9–12mm long; petals about twice as broad as sepals; lip 3-lobed; lip midlobe tapering and curved up towards column, crest with a torn blade 2–4mm high, arising transversely at the base of the lip; spur absent; column 5mm long, column foot 5mm long.

A S O N D J F M A M J J

Rare in shaded parts of riverine forests, bushveld and thickets, near sea level to 1,000m. Flowers between November and December.

Eulophia hereroensis

Ohrigstad 27.10.2010

Ohrigstad 30.10.2012

Ohrigstad 30.10.2012

Ohrigstad 27.10.2010

Rather slender terrestrial 250–550mm tall. **Leaves** thin-textured, partly to fully developed at flowering time, up to 450 × 20mm. **Inflorescence** lax and elongate, 7–25-flowered. **Flowers** with petals yellowish-green sometimes tinged with purple, petals and lip pale yellowish-green, crests pale lemon-yellow. Sepals 13–16mm long; petals slightly shorter and broader; lip 3-lobed, crests consisting of thin blades on all main veins, rather low on side lobes and tall on midlobe; spur flattened, 3–4mm long; column 6–8mm long, column foot 4–9mm long.

| A | S | O | N | D | J | F | M | A | M | J | J |

Scattered in arid vegetation, from near sea level to 1,700m. Flowers mainly between September and November.

Eulophia hians

Slender terrestrial 110–920mm tall. **Leaves** partly to fully developed at flowering time, up to 730mm long. **Inflorescence** lax, 3–40-flowered. **Flowers** with sepals purplish-green to dark reddish-purple, petals white tinged with pale pink to pale blue or yellow, sometimes purple-edged, crests pale pink to straw, sometimes tipped with dark purple. Sepals and petals slightly spreading; median sepal 8–18mm long; petals often united with the back of the column and curved or variously distorted; lip 3-lobed; lip side lobes often distorted and reduced, crests consisting of very low ridges in basal half, rising to 3–7 thin blades, variously toothed and papillose; spur slender, cylindrical to slightly club-shaped, 1.4–8.9mm long; column 4–7mm long, column foot absent to 2mm long.

Eulophia hians var. *hians*

Syn: *Eulophia clavicornis* var. *clavicornis*

Garden Castle 21.10.2012

White River 20.09.2010

Lydenburg 23.09.2010

Umtamvuna 29.09.2012

Umtamvuna 11.12.2011

KZN south coast 29.11.2013

Johannesburg 11.09.2012

Lydenburg 23.09.2010

Stutterheim 09.12.2013

Var. *hians* is distinguished as follows: leaves generally less than half the length of the flowering stem at flowering time; sepals dark reddish-purple; petals and lip white to pale pink or blue, sometimes purple-edged; lip often distorted, crests on lip midlobe consisting of toothed blades; lip side lobe veins gradually curving away from central axis of lip; spur 4–7mm long.

Distinguishable from *E. zeyheriana* by the thin papillose blades on the lip and the longer spur.

Var. *hians* occurs in Zimbabwe, South Africa, Swaziland and Lesotho.

Common in grassland, from near sea level to 2,200m. Self-pollinating. Flowers between August and December, and in the Drakensberg up to February.

Eulophia hians var. *inaequalis*

Syn: *Eulophia clavicornis* var. *inaequalis*; *Eulophia inaequalis*

Lydenburg 10.08.2012

Lydenburg 23.09.2011

Lydenburg 25.09.2010

Lydenburg 23.09.2010

Var. *inaequalis* is distinguished as follows: leaves partly developed at time of flowering; sepals green to dark reddish-brown, paler within; petals and lip bright yellow; crests on lip midlobe consisting of toothed blades; lip side lobe veins not curving away from central axis of lip.

Var. *inaequalis* occurs in Zimbabwe, South Africa, Swaziland and Lesotho.

Occasional in grassland, from near sea level to 2,000m. Self-pollinating. Flowers between August and November.

| A | S | O | N | D | J | F | M | A | M | J | J |

Eulophia hians var. *nutans*

Syn: *Eulophia clavicornis* var. *nutans*; *Eulophia nutans*

Garden Castle 03.02.2007

Warburton 09.01.2011

Garden Castle 03.02.2007

Var. *nutans* is distinguished as follows: leaves fully developed at time of flowering, 170–730mm long; petals and lip rarely curved or distorted; crests on lip midlobe consisting of numerous papillae, often joined at the base.

Var. *nutans* is widespread in Africa.

Common in grassland, sometimes in marshy conditions, from near sea level to 1,800m. Self-pollinating. Flowers between November and February.

| A | S | O | N | D | J | F | M | A | M | J | J |

Eulophia horsfallii

Syn: *Eulophia rosea*

Umtamvuna 27.01.2010

Eshowe 18.01.2013

Umtamvuna 27.01.2010

Mtunzini 18.01.2013

Very robust terrestrial 500mm–2m tall. **Leaves** tapering, 400mm–1.5m long. **Inflorescence** lax to fairly dense, 10–40-flowered. **Flowers** with sepals green tinged with brownish-purple, petals pale purple to white outside, paler within, purple-veined, lip green and purple, crests dull white. Sepals and petals spreading; sepals 15–22mm long; petals slightly longer than sepals; lip 3-lobed, crests consisting of 3–5 blades centrally, rising to more than 3mm on midlobe; spur 3–5mm long; column 9–16mm long, column foot 6–10mm long.

| A | S | O | N | D | J | F | M | A | M | J | J |

Sporadic in swampy soils, sometimes in the shade of bushes or forest trees, from near sea level to 1,000m. Pollinated by carpenter bees. Flowers between September and March.

Eulophia leachii

Khombaso 13.12.2012

Khombaso 14.12.2012

Khombaso 14.12.2012

Khombaso 11.12.2012

Khombaso 11.12.2012

Rather slender terrestrial 450–600mm tall; pseudobulbs 70–140mm long, mostly above ground. **Leaves** somewhat leathery, with rough margins, up to 200–300 × 8–15mm. **Inflorescence** lax, 3–24-flowered. **Flowers** with sepals yellowish-green often suffused with purple, petals yellowish-green with purple veins, lip greenish-yellow at the base to white at the tip. Sepals erect, 15–19mm long; petals similar but broader, apices recurved; lip obscurely 3-lobed, crests consisting of densely crowded papillae centrally, flanked by low ridges and occasional short papillae; spur 3–5mm long; column 9–11mm long, with a papilla 1.2mm long arising on either side of the rostellum, column foot 4–6 × 3–4mm.

A S O N D J F M A M J J

Distinguishable from *E. petersii* by the 4 or 5 soft, leathery leaves and smooth pseudobulbs.

Restricted but often in sizeable populations in bushveld under trees and in thorny thicket, from 100–1,000m. Flowers between December and January.

Eulophia livingstoneana

Tzaneen 30.10.2012

Tzaneen 30.10.2012

Tzaneen 30.10.2012

Tzaneen 30.10.2012

Slender terrestrial 300–700mm tall. **Leaves** thin-textured, absent or partly developed at flowering time. **Inflorescence** lax, 3–12-flowered. **Flowers** with sepals and petals white to pale pink, lip side lobes pale brown, midlobe purplish-pink. Sepals and petals spreading to reflexed, similar, 9–19 × 4–6mm; lip 3-lobed; lip side lobes erect, rounded, crests consisting of 2 broad, warty ridges; spur 4–5mm long; column 5–10mm long, column foot joined to lip side lobes.

A S **O** N **D** J F M A M J J

Occasional in *Brachystegia* woodland, from 800–1,200m. Flowers between October and December, but mainly in November.

Eulophia longisepala

Ingwavuma 30.11.2012

Ingwavuma 30.11.2012

Ingwavuma 04.12.2013

Ingwavuma 04.12.2013

Ingwavuma 04.12.2013

Slender terrestrial 220–630mm tall. **Leaves** thin-textured, absent to partly developed at flowering time, up to 200 × 6mm. **Inflorescence** lax, 2–10-flowered. **Flowers** with sepals green tinged with brownish-purple, petals green outside and yellow-green within, veins dull purple, lip side lobes pale yellow-green, midlobe white, crests pale green. Sepals erect, narrowly linear to spatula-shaped, 17–46mm long; petals about half as long; lip 3-lobed, crests consisting of 3 low ridges merging into a few rows of blades and papillae; spur 2–4mm; column 6–9mm long, column foot 2–5mm long.

A S O N D J F M A M J J

Very rare and localized in coastal sand and bushveld, from near sea level to 250m. Flowers between November and December.

Eulophia macowanii

Stutterheim 13.01.2011

Stutterheim 13.01.2011

Stutterheim 14.01.2011

Stutterheim 12.12.2013

Fairly stout terrestrial 230–400mm tall. **Leaves** partly to fully developed at flowering time, somewhat leathery, up to 150–400 × 10–30mm. **Inflorescence** rather lax, 4–20-flowered. **Flowers** mostly non-resupinate, sepals pale to dark chestnut-brown outside, pale green within, petals and lip white tinged with yellow, uncrested main veins purple, crests pale to bright yellow. Sepals and petals partly spreading; sepals 20–28mm long; petals shorter to a little longer than sepals, usually more or less circular; lip crests consisting of 2–5 blades in basal half, rising to 2mm, with a few rows of tapering papillae further on, usually terminating about halfway along midlobe; spur cylindrical, 5–8mm long; column 5–8mm long, column foot 1–2.5mm long.

A S O N D J F M A M J J

Distinguishable from *E. parvilabris* by the numerous slender papillae on the central third of the lip, generally narrower leaves and white lip with only the main lateral veins purple near the base.

Rare or locally frequent in grassland, from 30–1,200m. Flowers between November and February.

Eulophia meleagris

Luneburg (cultivated plant) date unknown

Luneburg (cultivated plant) date unknown

Locality unknown (cultivated plant) date unknown

Slender to fairly robust terrestrial 200–800mm tall. **Leaves** thin-textured, fully developed at flowering time, up to 300–600 × 20–55mm. **Inflorescence** lax, 3–30-flowered. **Flowers** with sepals dark lime-green on outer surface, dark brownish-purple with flecks of dull green within, petals and lip purple to white, with dark bluish-purple along margins and veins, crests pale purple basally, to white and dark purple near the tip. Sepals spreading, 14–17mm long; petals half the length of the sepals; lip 3-lobed; lip midlobe long and narrow, with 3–5 wavy blades, tallest on central veins in distal part of midlobe; spur 4–5mm long, often abruptly decurved apically; column 4–5mm long, column foot almost absent.

A S O N D J F M A M J J

Very rare and localized near margins of upland forests, often in dense herbaceous vegetation, from 1,200–2,000m. Flowers between December and January.

Eulophia ovalis

Robust terrestrial 150–650mm tall. **Leaves** somewhat leathery, fully developed and generally more than half the length of the flowering stem at flowering time, up to 110–620 × 4–30mm. **Inflorescence** lax to dense, 3–18-flowered. **Flowers** with sepals, petals, lip similar but with purplish veins on side lobes, crests white to yellow. Sepals spreading, 14–36mm long; petals bent over the lip to slightly spreading, about as long as sepals; lip 3-lobed, crests consisting of 2–5 wavy ridges in basal third and becoming slender, tapering papillae extending more or less halfway along midlobe; spur 1–5.6mm long; column 6–9mm long, column foot 1–5mm long.

Eulophia ovalis var. *bainesii*

Syn: *Eulophia bainesii; Eulophia ovalis* subsp. *bainesii*

Dullstroom 27.01.2011

Lydenburg 09.01.2011

Dullstroom 03.01.2014

Var. *bainesii* is distinguished as follows: leaves up to 620mm long; flowering stem up to 650mm tall; lip always cream-yellow; spur stout, flattened, 1–2.9mm long; crest papillae extending less than halfway along midlobe.

Var. *bainesii* occurs from Kenya southwards to South Africa and Swaziland.

Rare to scattered or locally frequent in open grassland or grassy places in bush- or thornveld, in dry to seasonally marshy situations, from 1,000–2,000m. Hybridizes with *E. zeyheriana*. Flowers between December and January.

A S O N D J F M A M J J

Eulophia ovalis var. *ovalis*

Machadodorp 09.01.2011

Ngeli 23.01.2010

Witsieshoek 14.01.2008

Garden Castle 23.01.2010

Var. *ovalis* is distinguished as follows: leaves up to 460mm long; flowering stem up to 500mm tall; lip white, yellow, pink or lilac; spur conical to cylindrical, 2.5–5.6mm long; crest papillae extending more than halfway along midlobe.

Var. *ovalis* occurs in Zimbabwe, South Africa, Swaziland and Lesotho.

Rare to scattered or locally frequent in open grassland, from 500–1,900m. Pollinated by halictid bees. Flowers mostly between November and February.

Eulophia parviflora

Fairly robust terrestrial 200–500mm tall. **Leaves** not yet, partly or fully extended at flowering time, leathery, up to 250 × 16mm. **Inflorescence** lax to rather dense, 5–30-flowered. **Flowers** resupinate or not, sepals dull brownish-green outside, mottled with rich orange-brown inside, petals yellow, veins marked red-brown inside, lip midlobe bright yellow, side lobes tinged with purple. Sepals and petals similar or petals a little shorter than median sepal and usually a little broader; sepals 10–14mm long; lip 3-lobed, crests consisting of low, broad, warty ridges on central veins; spur more or less cylindrical to shortly conical, 2–9mm long; column 4–8mm long, column foot usually a little shorter.

Eulophia parviflora (long-spurred form)

KZN south coast 27.06.2013

Kloof 19.07.2012

KZN south coast 24.07.2013

KZN south coast 24.07.2013

The long-spurred form is distinguished as follows: sterile shoot not yet developed at flowering time; inflorescence short, around 250mm, laxly few-flowered; flowers resupinate; spur 6–9mm long.

The long-spurred form occurs primarily at lower altitudes along the coast of KwaZulu-Natal but can also be found at higher altitudes in the mountains of Mpumalanga and Swaziland.

Occasional in grasslands, from near sea level to 1,800m. Pollinated by small long-tongued bees. Flowers between late June and September.

| A | S | O | N | D | J | F | M | A | M | J | J |

Eulophia parviflora (short-spurred form)

Lydenburg 23.09.2011

Lydenburg 25.09.2011

Peacevale 04.09.2012

Port Elizabeth 01.10.2004

The short-spurred form is distinguished as follows: sterile shoot partly to well developed at flowering time; inflorescence long, around 450mm, densely many-flowered; flowers non-resupinate; spur 2–5mm long.

The short-spurred form occurs mostly inland at higher altitudes through the Eastern Cape, KwaZulu-Natal and Mpumalanga, but can also be found near Durban.

Occasional but often locally abundant in grassland, sometimes growing in dense clumps, from near sea level to 1,800m. Pollinated by small flower chafer beetles. Flowers between September and December.

| A | S | O | N | D | J | F | M | A | M | J | J |

Eulophia parvilabris

Dullstroom 22.01.2012

Dullstroom 31.01.2012

Dullstroom 23.01.2012

Dullstroom 23.01.2012

Robust terrestrial 300–850mm tall. **Leaves** fully developed at flowering time, outer short and broad, inner up to 600 × 55mm. **Inflorescence** dense as the first flowers open, becoming lax and elongate, 6–18-flowered. **Flowers** non-resupinate, sepals pale purplish-brown, petals pale straw with purple bases, basal third of lip dark maroon, pale yellow near the tip. Sepals and petals spreading, similar, 27–31mm long; lip often shorter than petals, crests consisting of 2–7 laterally almost merging ridges on basal third, sometimes terminating in a short tooth; spur 5–6mm long; column 5–6mm long, column foot almost absent to 1mm long.

Distinguishable from *E. macowanii* by the crest consisting of merging ridges, the wider leaves and the coloration of the lip.

A S O N D J F M A M J J

Occasional and dispersed in open grassland, on moist steep slopes and flats, and at the base of rock faces, from 1,000–2,000m. Flowers between December and February.

Eulophia petersii

Karino 17.11.2012

Muden 05.01.2012

Lowveld Botanical Garden 19.01.2010

Lake Phobane 21.12.2013

Lake Phobane 17.12.2012

Robust terrestrial 900mm–2m tall; aerial pseudobulbs 60–150mm long, bearing leaves near the apex. **Leaves** 2–4, stiffly erect to curved, thick-textured, with rough margins and a sharp tip, up to 400 × 44mm. **Inflorescence** lax, 30–200-flowered, usually branched. **Flowers** with sepals dark green and purple, petals green, lip side lobes green, midlobe white tinged with pink. Sepals spreading, widely diverging, often rolled near apex, 23–28mm long; petals covering the column, slightly shorter than sepals, rolled near apex; lip 3-lobed, crests consisting of blades rising to more than 1.5mm tall centrally, often somewhat dissected; spur 2–8mm long; column 10–13mm long, column foot 2–8mm long.

A S O N D J F M A M J J

Distinguishable from *E. leachii* by the 2–4 hard, succulent leaves and ridged pseudobulbs.

Fairly common in sheltered to quite exposed places in valley bushveld and thorn savanna, often forming large clumps, from near sea level to 1,000m. Flowers between November and March.

Eulophia platypetala

Riversdale 24.11.2004

Joubertina 23.11.2011

Barrydale 30.11.2011

Slender to fairly robust terrestrial 250–450mm tall. **Leaves** somewhat leathery, partly developed at flowering time, up to 230 × 16mm. **Inflorescence** lax, 6–15-flowered. **Flowers** with sepals greenish-purple to chestnut-brown, paler on inner surface, petals and lip pale, slightly greenish lemon-yellow, main veins bluish-grey along margins of midlobe and on lip side lobes. Sepals partly spreading, 15–17mm long; petals about as long as median sepal but slightly wider; lip 3-lobed, crests consisting of 3–5 rather fleshy, almost entire blades, rising to 2mm high on midlobe and reaching to within 2mm of lip apex; spur 2–3mm long; column 7–11mm long, column foot 4–9mm long.

| A | S | O | N | D | J | F | M | A | M | J | J |

Rare in dry coastal renosterveld and fynbos, from near sea level to 800m. Flowers between October and November, stimulated by fire.

Eulophia schweinfurthii

Leydsdorp 19.11.2013

Leydsdorp 19.12.2013

Leydsdorp 19.11.2013

Leydsdorp 19.11.2013

Leydsdorp 20.12.2013

Robust terrestrial 500–980mm tall. **Leaves** slightly succulent, partly developed at flowering time, up to 350 × 10mm. **Inflorescence** lax, 12–25-flowered. **Flowers** with sepals brown to purplish-brown, petals clear yellow on outside, heavily veined to suffused with reddish-brown inside, raised part of lip bright yellow, side lobes with reddish-brown markings. Sepals 7–9mm long, reflexed; petals spreading, 12–14mm long; lip 3-lobed, crests consisting of bumpy ridges, mainly on convex portion of midlobe; midlobe abruptly bending inward, with the centre of the concavity opposite column apex; spur 3mm; column 5mm long, column foot 5–6mm long.

A S O N D J F M A M J J

Very rare in lowveld bush or inland mountain grassland, in unleached subtropical sandy soils, from 300–1,000m. Flowers between November and December.

Eulophia speciosa

Syn: *Eulophia austrooccidentalis; Eulophia leucantha; Eulophia wakefieldii*

Kenton-on-Sea 18.11.2010

Kenton-on-Sea 12.11.2011

Lowveld Botanical Garden 19.01.2010

Hluhluwe 05.10.2012

Sedgefield 05.12.2008

Robust terrestrial 400–900mm tall; partly aerial pseudobulbs sometimes present. **Leaves** succulent, fully developed at flowering time, up to 650 × 20mm. **Inflorescence** lax, 10–30-flowered. **Flowers** with sepals pale green, petals and lip yellow, lip side lobes with purple lines, crests deep yellow. Sepals reflexed, 11–13mm long; petals spreading to reflexed at apex, 10–22mm long; lip 3-lobed, crests of broad fleshy ridges, thickest on convex part of midlobe; spur 1–3mm long; column 5–8mm long, column foot 3–6mm long.

| A | S | O | N | D | J | F | M | A | M | J | J |

Common along the coast in open to rather sheltered places in bushveld, occasional from inland localities in valley bushveld to the thorny bush of the lowveld, and in mountain grassland, from near sea level to 900m. Pollinated by carpenter bees. Flowers between October and January.

Eulophia streptopetala

Richmond 17.11.2011

Sabie 13.01.2010

Sabie 13.01.2010

Lake Phobane 16.06.2012

Robust terrestrial 400mm–1.5m tall; partly aerial pseudobulbs joined at the base. **Leaves** thin-textured, up to 750 × 110mm. **Inflorescence** lax, 4–35-flowered. **Flowers** with sepals green variously mottled with dark purplish-brown, petals bright lemon-yellow, paler on inner surface, lip midlobe yellow, side lobes dull purple, crests deep yellow. Sepals widely spreading, 11–18mm long; petals as long as median sepal but wider, projecting over the column; lip 3-lobed, with crests consisting of low almost merging ridges on central veins; spur 1.5–2.2mm long; column 5–9mm long, column foot 5–9mm long.

Var. *streptopetala* is widespread in Africa and the only variety occurring in our region.

| A | S | O | N | D | J | F | M | A | M | J | J |

Common under the shelter of bushes, trees or large herbs, never in open grassland, sometimes in alien *Eucalyptus* plantations, from near sea level to 1,500m. Pollinated by leafcutter bees. Flowers between September and January.

Eulophia tenella

Port Alfred 16.12.2010

Port Alfred 17.12.2010

Dullstroom 31.01.2011

Port Alfred 16.12.2010

Slender terrestrial 150–200mm tall, rarely up to 600mm. **Leaves** fully developed at flowering time, up to 100–420 × 2–6mm. **Inflorescence** moderately dense, 5–25-flowered. **Flowers** with sepals dark green to brownish-purple, petals dull straw outside, pale brownish-purple within, lip dull pale purplish-brown, median crest on lip midlobe bright yellow. Sepals and petals partly spreading, similar; sepals 6–8mm long; petals as long as sepals but slightly wider; lip 3-lobed, crests consisting of fleshy almost entire ridges and blades, rising to approximately 0.7mm high on the 3 central veins; spur stoutly cylindrical, 2–3mm long; column 3–5mm long, column foot absent.

A S O N D J F M A M J J

Rare to locally frequent in areas with sourveld grassland and grassy places in coastal thornveld, from near sea level to 2,000m. Self-pollinating. Flowers between November and January.

Eulophia tuberculata

Grahamstown 17.11.2010

Grahamstown 27.11.2013

Grahamstown 27.11.2013

Grahamstown 23.11.2010

Slender terrestrial 150–400mm tall. **Leaves** succulent, widely spreading, absent to partly developed at flowering time, up to 240 × 18mm. **Inflorescence** lax, 10–30-flowered. **Flowers** with sepals yellowish-green heavily suffused with brown or purple, petals and lip bright yellow outside and pale yellow, heavily veined or suffused with reddish-purple inside, lip crests yellow. Sepals reflexed, 6–7mm long; petals spreading, wider than and 1.5 times as long as sepals; lip 3-lobed, crests on midlobe in basal half consisting of low fleshy ridges centrally, rising further on to finely toothed, fleshy blades terminating abruptly near midlobe apex; spur almost absent; column 3–4mm long, column foot 2–3mm long and with 2 yellow calli.

A S O N D J F M A M J J

Widespread in exposed places in dry grassland, karroid scrub or savanna, from 200–1,600m. Flowers between September and December.

Eulophia zeyheriana

Cathedral Peak 19.01.2008

Greytown 18.01.2012

Garden Castle 17.01.2014

Garden Castle 17.01.2014

Garden Castle 17.01.2014

Very slender terrestrial 150–350mm tall. **Leaves** stiffly erect, more than half the length of the flowering stem at flowering time, 200–400 × 3–7mm. **Inflorescence** lax, 4–17-flowered. **Flowers** with sepals green suffused with purplish-brown, petals pale blue with margins and apices tinged with purple, lip pale blue, crests mostly white. Sepals and petals partly spreading; sepals 7–9mm long; petals somewhat longer than sepals; lip 3-lobed, crests consisting of fleshy warty ridges on central veins of the lip, rising to approximately 0.5mm high on midlobe; spur 2–3mm long; column 3.5mm long, column foot absent.

Distinguishable from *E. hians* var. *hians* by the fleshy, warty ridges on the lip and by the shorter spur.

| A | S | O | N | D | J | F | M | A | M | J | J |

Local and scattered in high-altitude, open grassland, occasionally somewhat marshy, from 1,000–2,200m. Pollinated by halictid bees. Hybridizes with *E. ovalis* var. *bainesii*. Flowers between December and February.

Evotella rubiginosa

Franschhoek 07.12.2007

Franschhoek 06.12.2007

Porterville 09.12.2009

Franschhoek 10.12.2007

Slender terrestrial up to 330mm tall. **Leaves** 5–many, cauline, narrowly linear to spear-shaped, up to 90 × 6mm. **Inflorescence** dense, many-flowered; bracts about as long as ovary. **Flowers** with sepals green tinged with maroon, petals rusty red to maroon, lip white tinged with maroon, lip appendage green. Median sepal 6–7.5 × 2–3mm; lateral sepals spreading, 5–8 × 4–4.5mm, 2mm deep; petals fused to median sepal, more or less triangular, 5–7 × 5–6mm; lip broadly triangular, 5–7 × 6–9mm; appendage erect, 4–5 × 3mm, apically 2-lobed.

A S O N D J F M A M J J

Rare in marshy fynbos, from 20–1,300m. Flowers between October and December, after fire.

Gastrodia sesamoides

Cape Peninsula 30.11.2008

Cape Peninsula 30.11.2008

Cape Peninsula 22.11.2013

Cape Peninsula 22.11.2013

Cape Peninsula 30.11.2008

Slender achlorophyllous terrestrial 300mm–1.5m tall; stem brownish and stiff. **Leaves** reduced to membranous, cauline scales 5–10mm long. **Inflorescence** lax, up to 20-flowered. **Flowers** non-resupinate, projecting away or hanging, yellow to cream, bell-shaped. Sepals and petals similar, 12–15mm long, fused for most of their length; lip largely free, mobile, fused basally to the tube, obscurely 3-lobed, 10–12 × 5mm, with a callus of 2 crests.

A S O N D J F M A M J J

This is the only naturalized exotic orchid in southern Africa. Occasional and very localized in indigenous forest, plantation forest and fynbos, from 100–500m. Flowers between November and December.

Habenaria anguiceps

Naudesnek Pass 08.01.2009   Naudesnek Pass 05.02.2014   Naudesnek Pass 08.01.2009

Naudesnek Pass 04.02.2014   Naudesnek Pass 03.02.2014

Robust terrestrial up to 400mm tall. **Leaves** numerous along the stem, narrowly egg- to spear-shaped, up to 45 × 12mm. **Inflorescence** dense, 25–50-flowered. **Flowers** green, with petals yellowish-green. Median sepal held horizontally, 5mm long; lateral sepals deflexed, 6mm long; petals undivided, as long as median sepal and adherent to it; lip undivided, 7mm long; spur strongly inflated apically, 8mm long.

A S O N D J F M A M J J

Rare in well-drained grassland, from 200–2,500m. Flowers between November and February.

Habenaria arenaria

Nature's Valley 22.07.2012

Stutterheim 28.02.2011

Stutterheim 28.02.2011

Nature's Valley 24.06.2013

Port Alfred 24.06.2013

Nature's Valley 13.02.2012

Slender terrestrial up to 410mm tall. **Basal leaves** 2 or sometimes 3, not pressed to the ground, narrowly to broadly elliptic, occasionally mottled with grey, up to 170 × 50mm; **cauline leaves** much smaller. **Inflorescence** lax, 8–25-flowered. **Flowers** green. Median sepal erect, 3–4mm long; lateral sepals deflexed, 5mm long; petals undivided, 4mm long; lip 3-lobed; lip midlobe narrowly oblong, 6–7mm long; lip side lobes spreading, equal to or shorter than midlobe, 4–6mm long; spur 15–20mm long.

A S O N D J F M A M J J

Relatively common in shaded, seasonally dry habitats from coastal dune scrub to inland forest, from near sea level to 1,500m. Flowers between February and July.

Habenaria barbertoni

Heidelberg (Gauteng) 27.02.2012

Heidelberg (Gauteng) 02.03.2013

Heidelberg (Gauteng) 23.02.2013

Heidelberg (Gauteng) 23.02.2013

Heidelberg (Gauteng) 27.02.2012

Fairly robust terrestrial up to 400mm tall. **Leaves** 10, cauline, up to 110 × 29mm. **Inflorescence** lax, 3–13-flowered. **Flowers** with sepals green, petals and lip white. Sepals 7–11mm long, median erect, laterals spreading; petals divided, upper lobe elliptic, as long as median sepal and adherent to it, lower lobe shorter; lip 3-lobed, 11–17mm long, with lobes almost equal; spur inflated and slightly compressed in apical third, 30–40mm long.

Distinguishable from *H. transvaalensis* by the longer spur and the upper petal lobe that is longer than the lower one.

| A | S | O | N | D | J | F | M | A | M | J | J |

Rare in open woodland and on rocky hillsides, from 1,000–1,850m. Flowers between February and March.

Habenaria bicolor

Bronkhorstspruit 29.03.1981

Slender terrestrial up to 540mm tall. **Leaves** numerous, cauline, narrowly elliptic, 35–110 × 6–8mm. **Inflorescence** fairly lax, 12–35-flowered; lower bracts much longer than the flowers. **Flowers** green, petals and lip yellowish-green. Sepals 5.5mm long, median erect, laterals spreading; petals as long as and adherent to median sepal; lip 3-lobed, 6mm long; lip midlobe erect in front of column, less than 1.5 times the length of side lobes; spur thread-like, 7–11mm long.

Distinguishable from *H. laevigata* by the spreading leaves and sepals, and the spur that is shorter than the ovary.

A | S | O | N | D | J | F | M | A | M | J | J

Very rare in well-drained grassland, from 1,200–1,700m. Flowers between March and April.

Habenaria clavata

Mount Fletcher 08.02.2008

Mount Fletcher 08.02.2008

Mount Fletcher 08.02.2008

Dullstroom 01.02.2011

Dullstroom 31.01.2011

Giant's Castle 06.02.2009

Fairly robust terrestrial up to 800mm tall. **Leaves** numerous, cauline, elliptic to spear-shaped, up to 130 × 40mm. **Inflorescence** lax, 5–18-flowered. **Flowers** green, stigmatic arms white. Sepals 11–19mm long, median erect, laterals deflexed and rolled up lengthwise; petals divided, lower petal lobe linear, 25–40mm long, curved upward like a long horn, twice as long as upper lobe; lip 3-lobed, deflexed, 20–25mm long; lip side lobes shorter than midlobe; spur inflated apically, 30–50mm long.

Superficially similar to *H. cornuta*, but distinguishable by the conspicuous white stigmatic arms, longer spur and the lower petal lobes that do not cross or touch one another.

A S O N D J F M A M J J

Sporadic in well-drained grassland, from 600–2,000m. Flowers between December and February.

Habenaria cornuta

Qacha's Nek 24.02.2009

Garden Castle 25.02.2007

Garden Castle 27.02.2009

Giant's Castle 26.02.2007

Naudesnek Pass 11.02.2008

Fairly robust terrestrial up to 600mm tall. **Leaves** numerous, cauline, elliptic to narrowly spear-shaped, up to 100 × 45mm. **Inflorescence** rather dense, 8–20-flowered. **Flowers** green. Sepals 6–16mm long, median deflexed, laterals somewhat rolled up lengthwise; petals divided, lower lobe curving upward like a horn, 20–45mm long, upper lobe erect, linear, 6–15mm long; lip 3-lobed, 9–19mm long, with lobes almost equal; spur twisted once and inflated apically, 15–25mm long. Abnormal plants may occur in which the flowers, particularly the petals, are not properly developed.

Superficially similar to *H. clavata*, but distinguishable by the inconspicuous green stigmatic arms, shorter spur and the lower petal lobes that touch or cross one another.

| A | S | O | N | D | J | F | M | A | M | J | J |

Occasional in seasonally wet or marshy grassland, from 1,000–2,200m. Flowers between February and April.

Habenaria culveri

Lydenburg 22.02.2011

Lydenburg 22.03.2010

Lydenburg 23.02.2011

Lydenburg 22.02.2011

Lydenburg 22.02.2011

Slender terrestrial up to 400mm tall. **Leaves** 7–10, cauline, elliptic, 90–125 × 30–37mm, upper 2 or 3 smaller. **Inflorescence** lax, 9–25-flowered. **Flowers** green. Sepals 3–5mm long, median erect, laterals spreading; petals divided, upper lobe erect, as long as median sepal and adherent to it, lower lobe shorter than upper; lip 3-lobed, 5mm long; lip side lobes very short; spur 7–9mm long.

Distinguishable from *H. malacophylla* by the lower petal lobe that is much shorter than the upper and by the remnant side lobes of the lip.

A S O N D J F M A M J J

Rare in deep shade in subtropical forest, from 150–1,800m. Flowers between February and March.

Habenaria dives

Garden Castle 09.02.2008   Garden Castle 03.02.2007   KZN south coast 08.12.2012

KZN south coast 07.12.2012   Naudesnek Pass 12.02.2008

Slender to robust terrestrial up to 730mm tall. **Leaves** numerous, cauline, up to 240mm long. **Inflorescence** fairly lax, 20–70-flowered. **Flowers** white, variously suffused or tipped with green. Sepals 4–6.5mm long, deflexed; petals divided, lower lobe broadly spear-shaped, shorter than the oblong upper lobe; lip 3-lobed, 3–7mm long; lip side lobes half the length of midlobe; spur slightly inflated apically, 9–15mm long. Abnormal plants may occur in which the flowers, particularly the petals, are not properly developed.

Distinguishable from *H. falcicornis* by the denser inflorescence of smaller flowers with a shorter spur.

A S O N D J F M A M J J

Common in well-drained grassland, from 15–2,300m. Flowers between December and February.

Habenaria dregeana

Giant's Castle 26.02.2008

Dargle 04.03.2014

Glengarry 27.02.2009

Giant's Castle 26.02.2008

Dargle 04.03.2014

Slender terrestrial 150–300mm tall. **Basal leaves** 2, pressed flat on the ground, egg-shaped to circular or kidney-shaped, 25–45 × 30–65mm; **cauline leaves** 15–many, linear to spear-shaped. **Inflorescence** dense, 20–60-flowered. **Flowers** green. Sepals 4–7.5mm long, median erect, laterals spreading or deflexed; petals divided, with or without hairs, upper lobe erect, almost as long as median sepal and adherent to it, lower lobe curved upward, much shorter than upper; lip 3-lobed, 4–7.5mm long; lip side lobes shorter and narrower than midlobe; spur slightly inflated apically, 8–13mm long.

Distinguishable from *H. lithophila* by the shorter lip side lobes and spur with a wide mouth.

A S O N D J F M A M J J

Common on grassy slopes, from 300–2,000m. Flowers between January and April.

Habenaria epipactidea

Mount Fletcher 26.02.2011

Bronkhorstspruit 30.01.2014

Bronkhorstspruit 31.01.2014

Bronkhorstspruit 30.01.2014

Coombs 05.03.2011

Robust terrestrial up to 500mm tall. **Leaves** 8–12, cauline, egg- to spear-shaped, up to 150 × 40mm. **Inflorescence** dense, 6–40-flowered. **Flowers** green, petals greenish-white, lip white. Median sepal erect, 8–12mm long, lateral sepals deflexed or spreading, longer and narrower than median sepal; petals undivided, rounded or square, 9–15mm long; lip midlobe hanging, strap-like, 11–16mm long; lip side lobes reduced, thread-like; spur knobbed apically, 20–30mm long.

This species is rather variable in stature and leaf shape and plants from the more arid areas are more slender and lax, with narrower leaves.

A S O N D J F M A M J J

Common in coastal and inland savanna, seasonally damp or marshy grasslands and on rocky slopes, from near sea level to 2,400m. Pollinated by hawkmoths. Flowers between January and early March.

Habenaria falcicornis

Fairly robust to robust terrestrial up to 830mm tall. **Leaves** numerous, cauline, spear-shaped, up to 80–180mm long. **Inflorescence** lax, 20–60-flowered. **Flowers** white, tipped with green. Sepals 4.5–8mm long, deflexed; petals divided, lower lobe spreading, broadly egg- to spear-shaped, upper lobe erect, narrowly elliptic to linear; lip 3-lobed, 7–12mm long; lip midlobe longer and narrower than side lobes; spur slightly inflated in apical half, 23–43mm long.

Habenaria falcicornis subsp. *caffra*

Syn: *Habenaria caffra; Habenaria falcicornis* var. *caffra*

Barberton 17.02.2010

Barberton 17.02.2010

Naudesnek Pass 22.02.2009

Cathedral Peak 08.02.2009

Subsp. *caffra* is distinguished as follows: habit more robust; flowers larger; lower petal lobe spear-shaped, as long as or longer than the upper lobe, 5–8mm long.

Subsp. *caffra* occurs from KwaZulu-Natal and Lesotho northwards, including Swaziland and Zimbabwe.

Fairly common in grassland, from 15–2,400m. Flowers between January and April.

| A | S | O | N | D | J | F | M | A | M | J | J |

Habenaria falcicornis subsp. *falcicornis*

Syn: *Habenaria falcicornis* var. *falcicornis*

Maclear 24.01.2010

Maclear 13.02.2008

Stutterheim 28.02.2011

Maclear 12.02.2008

Subsp. *falcicornis* is distinguished as follows: habit less robust; flowers smaller; lower petal lobe egg-shaped, shorter than the upper lobe, 2.5–5mm long.

Subsp. *falcicornis* occurs between Swellendam and southern KwaZulu-Natal and Lesotho.

Distinguishable from *H. dives* by the lax inflorescence of larger flowers with a longer spur.

Fairly common in seasonally marshy grassland, from 60–2,400m. Flowers between January and March.

A	S	O	N	D	J	F	M	A	M	J	J

Habenaria filicornis

Warburton 26.02.2013

Dullstroom 15.02.2010

Loteni 25.02.2008

Giant's Castle 27.02.2008

Loteni 10.02.2009

Slender terrestrial up to 600mm tall. **Leaves** few, cauline, linear to spear-shaped, up to 140 × 14mm. **Inflorescence** lax, 12–25-flowered. **Flowers** green. Median sepal erect, 3–4.5mm long, lateral sepals deflexed, 3.5–6mm long; petals undivided, as long as median sepal and adherent to it; lip 3-lobed; lip midlobe slightly longer and broader than side lobes, 5–8mm long; spur slightly inflated apically, 20–30mm long.

A S O N D **J F M** A M J J

Fairly common but inconspicuous in seasonally damp or marshy grasslands or floodplains, from 400–2,000m. Flowers between January and March.

Habenaria galpinii

Graskop 17.02.2010

Graskop 21.02.2011

Graskop 17.02.2010

Graskop 17.02.2010

Slender terrestrial up to 400mm tall. **Leaves** 5–14, clustered near the base, oblong to linear, 80–200 × 7–18mm. **Inflorescence** fairly lax, 16–35-flowered. **Flowers** green and white. Sepals glandular-hairy, reflexed, 5–8mm long; petals divided, shortly hairy, upper lobe erect, linear, 5–6mm long, lower lobe linear, curved, 8–12mm long; lip 3-lobed; lip midlobe 10–14mm long; lip side lobes 8–10mm long; spur inflated in apical half, 20–25mm long.

Distinguishable from *H. nyikana* by the leaves clustered near the base and by plant height, which is generally well below 400mm.

| A | S | O | N | D | J | F | M | A | M | J | J |

Fairly common in damp grassland, often along streams or on stony hillsides, from 900–2,000m. Flowers between February and April.

Habenaria humilior

Loteni 02.02.2009

Dullstroom 12.02.2010

Dullstroom 12.02.2010

Dullstroom 12.02.2010

Fairly robust terrestrial up to 600mm tall. **Leaves** numerous, cauline, 60–230 × 10–15mm. **Inflorescence** lax to moderately dense, 7–18-flowered. **Flowers** green with white centre. Sepals reflexed, 5–9mm long; petals divided, upper lobe recurved, linear, 5mm long, lower lobe elliptic-oblong or narrowly oblong, 8–9mm long; lip 3-lobed, deflexed; lip midlobe 10–11mm long; lip side lobes 7mm long; spur twisted through 180°, inflated in apical half, 15–20mm long; stigmatic arms and anther canals markedly divergent.

A S O N D J F M A M J J

Localized in damp grassland, from 1,200–2,050m. Flowers between January and February.

Habenaria kraenzliniana

Heidelberg (Gauteng) 27.02.2012

Heidelberg (Gauteng) 27.02.2012

Johannesburg 27.01.2011

Heidelberg (Gauteng) 27.02.2012

Slender terrestrial up to 370mm tall. **Basal leaves** 2, pressed flat on the ground, egg-shaped to circular or kidney-shaped, 25–45 × 35–46mm; **cauline leaves** 7–17, spear-shaped, up to 25mm long. **Inflorescence** lax, 10–25-flowered. **Flowers** light green. Sepals 6–8mm long, median erect, laterals spreading; petals divided, hairy, upper lobe spear-shaped, curved, as long as median sepal and adherent to it, lower lobe thread-like, much longer than upper, 30–45mm long; lip 3-lobed, shortly hairy; lip midlobe 11–13mm long; lip side lobes 23–38mm long; spur inflated in apical third, with a notched flap in front of the mouth, 35–45mm long.

A S O N D **J F M A M** J J

Rare on stony, grassy hillsides, from 1,000–1,800m. Flowers between the end of January and April.

Habenaria laevigata

Witsieshoek 17.12.2008

Garden Castle 16.01.2014

Ngeli 03.02.2013

Ngeli 03.02.2013

Slender to fairly robust terrestrial up to 400mm tall. **Leaves** numerous, clasping the stem, elliptic, margins with a horny edge, 30–60 × 7–15mm. **Inflorescence** fairly lax, 12–30-flowered; bracts much longer than and partly covering the flowers. **Flowers** green, petals and lip yellowish-green. Sepals 5–9.5mm long, median erect, laterals deflexed; petals undivided, 5–8.5mm long, adhering to median sepal; lip 3-lobed, 6–9.5mm long; lip midlobe erect in front of column, at least 1.5 times as long as side lobes; spur 16–35mm long.

Distinguishable from *H. bicolor* by the clasping leaves, deflexed sepals and spur that is longer than the ovary.

A	S	O	N	D	J	F	M	A	M	J	J

Occasional in well-drained, often stony grassland, from 660–2,200m. Flowers between December and February.

Habenaria lithophila

George 12.12.2010

Witsieshoek 15.01.2008

Sabie 24.01.2014

KZN south coast 23.09.2013

Sabie 25.01.2013

Naudesnek Pass 12.02.2008

Slender terrestrial up to 300mm tall. **Basal leaves** 2, pressed flat on the ground, egg-shaped to circular or kidney-shaped, 50 × 55–70mm; **cauline leaves** 5–30, spear-shaped. **Inflorescence** dense, 1–30-flowered. **Flowers** green. Sepals 5–8.5mm long, median erect, laterals spreading; petals divided, with or without hairs, upper lobe erect, adherent to median sepal, lower lobe curving upwards, slightly longer than upper lobe; lip 3-lobed, 6–9mm long, lobes almost equal in length, without hairs or shortly hairy; spur inflated apically, 8–11mm long.

Distinguishable from *H. tysonii* and *H. mossii* by more or less erect flowers, the erect median sepal concealing the column, flat upper petal lobes, lower petal lobes that are longer than the upper lobes and curved upward, and by the lip side lobes almost equalling the midlobe. Distinguishable from *H. dregeana* by the longer lip side lobes and narrow mouth of the spur.

A S O N D J F M A M J J

Common in grassland, from 300–2,500m. Flowers between December and March.

Habenaria malacophylla

Cathedral Peak 29.02.2008

Roayal Natal National Park 22.02.2010

Roayal Natal National Park 22.02.2010

Harding 27.02.2010

Ngeli 24.02.2009

Slender terrestrial up to 700mm tall. **Leaves** 8–15, cauline, mostly in the centre, with lower part of the stem rather bare, 80–160 × 20–37mm. **Inflorescence** lax, 12–30-flowered. **Flowers** green. Sepals 3.5–6mm long, median erect, laterals deflexed; petals divided, upper lobe as long as median sepal and adherent to it, lower lobe ascending, slightly longer and narrower than the upper; lip 3-lobed, 4–6.5mm long; lip midlobe broader and usually shorter than side lobes; spur somewhat thickened in the middle, then tapering, 8–15mm long.

Var. *malacophylla* is widespread in Africa and is the only variety in the region.

Distinguishable from *H. culveri* by the lower petal lobe that is slightly longer than the upper lobes, and the lip lobes almost equal to each other.

A S O N D **J F M A** M J J

Occasional on forest margins and forest floors, from 150–2,000m. Hybridizes with *H. tridens*. Flowers between January and April.

Habenaria mossii

Irene 16.03.2014

Irene 16.03.2014

Irene 16.03.2014

Irene 16.03.2014

Slender terrestrial up to 240mm tall. **Basal leaves** 2, broadly egg-shaped to circular, 35–70 × 30–70mm; **cauline leaves** 13–20, spear-shaped. **Inflorescence** fairly dense, many-flowered. **Flowers** green. Sepals 5–9mm long, median reflexed, laterals spreading; petals divided, hairy, upper lobe erect, adherent to median sepal, 5–7mm long, lower lobe upcurved, 5–9mm long; lip 3-lobed, hairy; lip midlobe 6–11mm long; lip side lobes 7–10mm long; spur twisted, slightly inflated in apical third, 18–26mm long.

Distinguishable from *H. tysonii* and *H. lithophila* by the spreading flowers with larger stigmas, reduced auricles and a much longer, twisted spur.

A S O N D J F M A M J J

Rare in grassland, on dolerite or black sandy soil, from 1,200–1,800m. Flowers between March and April.

Habenaria nyikana

Rosehaugh 24.03.2010

Rosehaugh 24.03.2010

Rosehaugh 24.03.2010

Rosehaugh 24.03.2010

Robust terrestrial up to 1.2m tall. **Leaves** 6–10, cauline, linear to spear-shaped, up to 350 × 20mm. **Inflorescence** lax to semi-dense, 18–40-flowered. **Flowers** green, centre white. Sepals deflexed, 5–10.5mm long; petals divided, upper lobe recurved, linear, 5.5–8mm long, lower lobe 9.5–13.5mm long; lip 3-lobed; lip midlobe 11–15mm long; lip side lobes 10–12.5mm long; spur twisted through 180°, distinctly inflated in apical half, 20–30mm long. Abnormal flowers with underdeveloped lower petal lobes and side lobes of the lip have been recorded.

Subsp. *nyikana* also occurs in Malawi, Zimbabwe, Mozambique and Swaziland.

Distinguishable from *H. galpinii* by the leaves borne along the stem and by plant height, which is generally above 500mm.

A S O N D J F M A M J J

Fairly common in damp grassland, from 600–1,700m. Flowers between February and April.

Habenaria petitiana

Lydenburg 17.02.2011

Lydenburg 17.02.2011

Lydenburg 22.03.2010

Lydenburg 17.02.2011

Slender terrestrial up to 350mm tall. **Leaves** 7–14, cauline, widely spaced, up to 80 × 40mm, the uppermost much smaller. **Inflorescence** lax, 6–40-flowered. **Flowers** green. Sepals 2.5–4mm long, median erect, laterals spreading; petals undivided, adherent to median sepal, 2–3.5mm long; lip 2.5–4mm long and nearly as broad, 3-lobed in apical half; lip side lobes often longer than midlobe; spur more or less spherical, 1mm long.

A S O N D J F M A M J J

Rare in shade of forest margins, from 1,200–1,850m. Flowers in February.

Habenaria pseudociliosa

Dullstroom 13.02.2011

Heidelberg (Gauteng) 15.03.2014

Dullstroom 21.02.2010

Heidelberg (Gauteng) 15.03.2014

Port Edward 19.12.2011

Slender to fairly robust terrestrial up to 500mm tall. **Leaves** 6–8, cauline, spear-shaped, up to 150 × 15mm; basal leaf sheaths with conspicuous horizontal black bars. **Inflorescence** fairly dense, 18–40-flowered. **Flowers** green. Sepals bristly along margins and veins, median erect, laterals spreading, 3–5mm long; petals undivided, as long as median sepal and adherent to it; lip 3-lobed, 4–5mm long; lip side lobes shorter than midlobe; spur inflated apically, 16–23mm long.

Distinguishable from *H. ciliosa* (not included in this guide) by the longer spur, at least 1.5 times longer than the ovary, and the lip with side lobes shorter than the midlobe.

Occasional in damp grassland, from near sea level to 1,800m. Flowers between January and March.

Habenaria schimperiana

Naudesnek Pass 07.02.2008

Naudesnek Pass 07.02.2008

Naudesnek Pass 07.02.2008

Naudesnek Pass 07.02.2008

Robust terrestrial up to 1m tall. **Leaves** 6–10, cauline, linear to spear-shaped, up to 280 × 20mm. **Inflorescence** lax, 10–25-flowered; flower stalk horizontal, 15mm long, ovary 10mm long, sharply turned downward, giving the flowers a dangling posture. **Flowers** green, centre white. Median sepal deflexed, 6–8mm long, lateral sepals deflexed and twisted, 9–11mm long; petals divided, lower lobe spreading downward, 14–18.5mm long, twice as long as the deflexed, linear upper lobe; lip 3-lobed, deflexed; lip midlobe 13–17mm long; lip side lobes shorter; spur with a knee-bend and twisted, inflated apically, 10–15mm long.

| A | S | O | N | D | J | F | M | A | M | J | J |

Localized in swampy or wet grassland, on poorly drained soil, from 1,000–2,200m. Flowers between December and February.

Habenaria transvaalensis

Syn: *Bonatea bracteata; Bonatea liparophylla*

Luneburg 11.02.2014

Luneburg 06.02.2012

Luneburg 11.02.2014

Luneburg 11.02.2014

Moderately robust terrestrial up to 550mm tall. **Leaves** numerous, cauline, elliptic, up to 120 × 40mm. **Inflorescence** fairly lax, 15–25-flowered. **Flowers** with sepals green, petals and lip white. Sepals 10–12mm long, median erect, laterals spreading; petals divided, lower petal lobe narrower than upper; lip 3-lobed, 10–12mm long, with lobes almost equal; spur somewhat inflated and compressed in apical half, 20mm long.

Distinguishable from *H. barbertoni* by the shorter spur and the upper petal lobe that is almost as long as the lower one.

A S O N D J F M A M J J

Rare in grassland, from 1,200–2,000m. Flowers between January and February.

Habenaria tridens

Roossenekal 05.02.2013

Roossenekal 08.01.2013

Roossenekal 08.01.2013

Roossenekal 13.01.2013

Slender terrestrial 180–320mm tall. **Leaves** 6–12, cauline, narrowly oblong, 80–100 × 10–20mm. **Inflorescence** lax, 10–30-flowered. **Flowers** green. Median sepal erect, 3.5–4.5mm long, lateral sepals deflexed, 4–5mm long; petals divided, upper lobe erect, as long as median sepal, lower lobe spreading or recurved, linear, as long as upper; lip 3-lobed, 5mm long; lip side lobes slightly narrower and shorter than midlobe; spur somewhat inflated apically, 9–15mm long.

A S O N D J F M A M J J

Occasional but inconspicuous along stream banks, often with the roots submerged, from 400–1,800m. Hybridizes with *H. malacophylla*. Flowers between January and March.

Habenaria trilobulata

KZN north coast 21.03.2013

KZN north coast 11.03.2013

KZN north coast 21.03.2013

KZN north coast 11.03.2013

KZN north coast 16.01.2013

Slender terrestrial up to 350mm tall. **Basal leaves** 2, pressed flat on the ground, elliptic to circular, silvery green with dark venation, 25–75 × 20–65mm; **cauline leaves** 5–8, bract-like. **Inflorescence** lax, 7–15-flowered. **Flowers** greenish-white, petals and spur white. Sepals 4.5–7mm long, median erect, laterals spreading; petals 2-lobed, upper lobe 3–4mm long, lower lobe spreading, 17–23mm long; lip 3-lobed; lip midlobe 6.5–8mm long; lip side lobes 20–27mm long; spur inflated apically, 18–30mm long.

| A | S | O | N | D | J | F | M | A | M | J | J |

Rare in open woodland, lowland and dry evergreen forest, from near sea level to 1,100m. Flowers between February and April.

Habenaria tysonii

Verloren Vallei 03.01.2011

Giant's Castle 20.01.2014

Giant's Castle 19.01.2014

Giant's Castle 20.01.2014

Giant's Castle 20.01.2014

Slender terrestrial up to 380mm tall. **Basal leaves** 2, pressed flat on the ground, egg-shaped to circular or kidney-shaped, 23–45 × 20–70mm; **cauline leaves** 9–many, spear-shaped, up to 28mm long. **Inflorescence** lax, 20–50-flowered. **Flowers** green. Sepals 6–9mm long, median partly reflexed, exposing the column, laterals spreading; petals divided, hairy, upper lobe somewhat shorter than median sepal and adherent to it, lower lobe twice as long as the upper; lip 3-lobed, hairy, 9–14mm long; lip midlobe slightly shorter than side lobes; spur inflated apically, often twisted, 12–20mm long.

Distinguishable from *H. lithophila* and *H. mossii* by the horizontally-held flowers with the narrow median sepal reflexed and thus exposing the column, by the twisted upper petal lobes, longer lower petal lobes and the lip side lobes that are curved backward.

Occasional on damp grassy, rocky slopes, from near sea level to 2,150m. Flowers between January and February.

Habenaria woodii

Umtamvuna 13.03.2010   Umtamvuna 03.03.2013   Umtamvuna 13.03.2010

Umtamvuna 13.03.2010   Umtamvuna 13.03.2010

Slender terrestrial up to 580mm tall. **Leaves** few, cauline, linear, 100–270 × 3mm. **Inflorescence** lax, many-flowered. **Flowers** green. Median sepal deflexed, 3mm long, lateral sepals, deflexed 4mm long; petals divided, upper lobe linear, shortly hairy near the base, 3mm long, lower lobe linear to spear-shaped, curved, shortly hairy basally, 4.5mm long; lip 3-lobed, lobes 5–6.5mm long, with side lobes somewhat curved and slightly shorter; spur inflated in apical third, 20mm long.

A S O N D J F M A M J J

Rare in marshy to moist subtropical grassland, from near sea level to 600m. Flowers between March and April.

Holothrix aspera

Citrusdal 11.08.2013

Ceres 02.09.2010

Worcester 20.08.2011

Worcester 15.08.2011

Slender terrestrial up to 250mm tall. **Leaves** 2, pressed flat on the ground, without hairs, up to 30mm broad. **Inflorescence** lax, more or less 1-sided, 3–24-flowered; stalk without bracts, sparsely to moderately covered with short soft hairs. **Flowers** white, centre green, petals and lip with purple markings. Sepals 1–3.5 × 0.8–2.5mm; petals 3.5–6.5 × 1–4mm; lip unequally 5–7-lobed, 3–8.5 × 1–6mm; spur strongly curved, 3–5mm long.

Distinguishable from *H. mundii* by the lax inflorescence, larger flowers and the spur curved abruptly forward.

A S O N D J F M A M J J

Common in semi-arid areas, in sandy or rocky soil or rock crevices, from 300–1,000m. Flowers between June and October.

Holothrix brevipetala

Houwhoek 19.11.2010

Grabouw 20.12.2009

George 18.01.2011

Cape Peninsula 21.12.2007

George 17.10.2011

Slender terrestrial up to 310mm tall. **Leaf** 1, pressed flat on the ground, densely covered with stout hairs, up to 12mm broad, mostly withered at flowering time. **Inflorescence** lax, more or less 1-sided, up to 35-flowered; stalk stout, without bracts, with stiff deflexed hairs. **Flowers** greenish-yellow. Sepals densely hairy, 1–2.5 × 0.5–1.5mm; petals 2.5–3.5 × 0.5–1.5mm; lip with 3–5 short, broad lobes of 2.5–3.5 × 2–3.5mm; spur slightly curved, 1–2mm long.

Distinguishable from *H. cernua* by the much shorter, broader lip lobes.

A	S	O	N	D	J	F	M	A	M	J	J

Occasional in sand and shallow soil in rock crevices, from near sea level to 1,100m. Flowers between October and January, mostly after fire.

Holothrix burchellii

Riversdale 16.10.2011   Krakeel 21.10.2010   Riversdale 04.10.2010

Riversdale 05.10.2010   Riversdale 05.10.2010   Riversdale 20.09.2011

Slender terrestrial up to 500mm tall. **Leaves** 2, pressed flat on the ground, without hairs, up to 60mm broad. **Inflorescence** dense, more or less 1-sided, up to 64-flowered; stalk with 2–7 bracts and reflexed hairs. **Flowers** cream-coloured, of two kinds. Sepals 2.5–5 × 1–2.5mm; petals 3–18 × 1–3.5mm, divided into 5–10 thread-like lobes, these much longer on upper than on lower flowers; lip 3–9 × 1.5–4mm, divided into 5–13 slender lobes, lobes longer on upper than lower flowers; spur strongly curved, smaller on upper than lower flowers, 3–6mm long.

A S O N D J F M A M J J

Occasional in a variety of habitats including fynbos, grassland and dry rocky slopes, from 30–1,200m. Flowers between August and November.

Holothrix cernua

Cape Peninsula 09.11.2008

Genadendal 29.09.2010

Franschhoek 18.10.2013

Franschhoek 18.10.2013

Paarl 25.06.2012

Slender terrestrial up to 240mm tall. **Leaves** 2, pressed flat on the ground, usually covered with short, stiff, hooked hairs, up to 28mm broad, sometimes withered at flowering time. **Inflorescence** lax to dense, more or less 1-sided, 28–45-flowered; stalk without bracts, with long deflexed hairs. **Flowers** cream to lime-green. Sepals hairy, 1.5–3 × 0.5–1.5mm; petals 3.5–7 × 0.5–1.5mm; lip with 3–5(–7) lobes; lip midlobe the longest, lobes comprising one third to half the length of the whole lip; spur curved, 1.5–4mm long.

Distinguishable from *H. brevipetala* by the petals and lip lobes that resemble a human hand, and from *H. villosa* by the deflexed hairs on the inflorescence stalk.

A S O N D J F M A M J J

Frequent in sandy or stony places, from near sea level to 1,500m. Flowers between July and January, mostly after fire.

Holothrix exilis

Riversdale 12.11.2010

Riversdale 22.10.2010

George 16.01.2011

George 18.01.2011

Riversdale 22.10.2010

Riversdale 30.10.2010

Very slender terrestrial up to 300mm tall. **Leaves** 2, thinly hairy above, up to 8mm broad, sometimes withered at flowering time. **Inflorescence** lax, more or less 1-sided, up to 30-flowered; stalk without bracts, densely to very sparsely hairy, with long fine hairs at right angles. **Flowers** creamy green. Sepals 0.8–3.5 × 0.5–1mm; petals 1.5–4 × 0.3–1mm; lip entire or 3-lobed, 1.8–3.5 × 0.5mm, outer lobes very short or up to about half the length of the midlobe; spur slightly curved, 0.8–1.8mm long.

   Distinguishable from *H. villosa* by the shorter spur and lip side lobes that are absent or very short.

A S O N D J F M A M J J

Occasional in sandy fynbos, from near sea level to 1,500m. Flowers between October and March, stimulated by fire.

Holothrix filicornis

Port Nolloth 26.05.2013

Port Nolloth 19.05.2012

Port Nolloth 19.05.2012

Port Nolloth 19.05.2012

Slender terrestrial up to 260mm tall. **Leaves** 2, pressed flat on the ground, without hairs, up to 40mm broad. **Inflorescence** lax, more or less 1-sided, up to 30-flowered; stalk without bracts, without hairs. **Flowers** with sepals green tinged with red, petals and lip white to greenish-white. Sepals 1.5–2 × 0.5–1mm; petals 3-lobed, 3–8 × 0.5–1mm; lip 5-lobed with lobes comprising up to three quarters of total lip length, 4.5–9 × 1.5–2.5mm; spur slightly curved, 7–11mm long.

| A | S | O | N | D | J | F | M | A | M | J | J |

Occasional in semi-arid areas, in rock crevices and on stony slopes, from 250–1,000m. Flowers between May and September.

Holothrix grandiflora

Elands Bay 02.04.2011

Elands Bay 09.03.2011

Elands Bay 09.03.2011

Oudtshoorn 18.03.2013

Fairly robust terrestrial up to 580mm tall. **Leaves** 2, pressed flat on the ground, without hairs, up to 80mm broad, often withered at flowering time. **Inflorescence** lax, more or less 1-sided, up to 40-flowered; stalk with 8–12 bracts, without hairs or with a few minute hairs at the base. **Flowers** white to pale green, often tinged with lilac. Sepals 4.5–9 × 1.5–4.5mm; petals divided into 5–9 thread-like lobes, 10–22 × 1.5mm; lip divided into 13–26 thread-like lobes, 12.5–25.5 × 3.5–6.5mm; spur slightly curved, 3–4.5mm long.

Distinguishable from *H. schlechteriana* by the shorter spur in relation to the lip, longer petals, nearly hairless inflorescence stalk and later flowering time.

A S O N D J F M A M J J

Rare in semi-arid areas, in rock crevices or under short scrub, from near sea level to 600m. Flowers between March and April.

Holothrix incurva

Sentinel 15.01.2008 Naudesnek Pass 03.02.2014 Naudesnek Pass 03.02.2014

Naudesnek Pass 03.02.2014 Naudesnek Pass 03.02.2014

Slender terrestrial up to 170mm tall. **Leaves** 2, with hairs on margins only, up to 30mm broad. **Inflorescence** lax, 1-sided, up to 20-flowered; stalk without bracts, with short, fine hairs at right angles. **Flowers** yellow to greenish-yellow. Sepals hairy, 2–4.5 × 1.5–2mm; petals 5–9.5 × 0.5–1mm; lip with 5 narrowly linear lobes, 3.5–4.5 × 1.5–2mm; spur curved, 1–1.5mm long.

Distinguishable from *H. thodei* by the presence of leaves at flowering time.

| A | S | O | N | D | J | F | M | A | M | J | J |

Frequent on basalt ledges and in rock crevices, from 1,500–2,900m. Flowers between December and April.

Holothrix longicornu

Cape Peninsula 29.10.2008

Cape Peninsula 29.10.2008

Cape Peninsula 29.10.2008

Cape Peninsula 26.10.2004

Cape Peninsula 11.09.2008

Slender terrestrial up to 160mm tall. **Leaves** 2, pressed flat on the ground, with dense, short, soft hairs, usually prominently veined, up to 20mm broad. **Inflorescence** lax, more or less 1-sided, up to 28-flowered; stalk without bracts and with deflexed hairs. **Flowers** yellow-green. Sepals 1.5–2.5 × 0.5mm; lip divided into 3–5 short, broadly linear lobes, 2.5–3 × 1–1.5mm; spur longer than lip, 3–4mm long.

A S O N D J F M A M J J

Very rare in fynbos, from near sea level to 250m. Flowers in October.

Holothrix majubensis

Volksrust 15.01.2012

Volksrust 15.01.2012

Volksrust 20.12.2011

Volksrust 20.12.2011

Slender terrestrial up to 55mm tall. **Leaves** 2, pressed flat on the ground, hairy on margins only, up to 35mm broad. **Inflorescence** fairly dense, 1-sided, up to 25-flowered; stalk without bracts, hairy. **Flowers** non-resupinate, white. Sepals hairy, 2 × 1.2mm; petals 2.8 × 1mm; lip with 3 almost equal lobes, 2.5–3 × 1–2mm; spur recurved, 1mm long.

A S O N D J F M A M J J

Known only from one locality where it is locally common in cracks in vertical sandstone cliffs under *Merwilla plumbea* (syn: *Scilla natalensis*), from 2,000–2,225m. Flowers between December and January.

Holothrix mundii

Genadendal 21.09.2011

Riversdale 16.09.2013

Genadendal 23.08.2010

Genadendal 28.09.2010

Slender terrestrial up to 160mm tall. **Leaves** 2, pressed flat on the ground, with short hairs on margins only, up to 20mm broad. **Inflorescence** condensed, 3–18-flowered; stalk without bracts, with short, slightly reflexed hairs. **Flowers** white, anther compartments bright pink-mauve or purple. Sepals without hairs, 1–2 × 0.3–0.5mm; petals 2mm long; lip 3–7-lobed, 1.5–3.5 × 0.5–2mm; lip midlobe and outer pair longer than other 4 lobes; spur straight, 0.5–1.5mm long.

Distinguishable from *H. aspera* by the condensed inflorescence, smaller flowers and straight spur.

A S O N D J F M A M J J

Occasional but inconspicuous in fynbos and grassland, from sea level to 1,100m. Flowers between September and November.

Holothrix orthoceras

Giant's Castle 03.02.2009

Giant's Castle 03.02.2009

Sabie 09.04.2013

Giant's Castle 16.12.2008

Sabie 09.04.2013

Slender terrestrial, epiphyte or lithophyte up to 280mm tall. **Leaves** 2, not quite flat on the ground, without hairs, with white or silvery net-veins, up to 40mm broad. **Inflorescence** lax, more or less 1-sided, 8–30-flowered; stalk without bracts, densely covered with short, soft hairs. **Flowers** white, lip with lilac-blue veins. Sepals 1.5–3.5 × 0.5–1mm; petals rolled into a tube, 3.5–7.5 × 1–2mm; lip divided into 3 short front lobes and 2 longer side lobes, 4–8.5 × 2–5.5mm; spur straight, 2.5–6.5mm long.

Distinguishable from *H. parviflora* by the straight spur and triangular rather than rectangular lobes in the centre of the lip.

A S O N D J F M A M J J

Common on mossy rocks in forests, from 300–1,900m. Flowers between January and June.

Holothrix parviflora

Wilderness 23.06.2013

Karatara 03.10.2010

Karatara 03.10.2010

Slender terrestrial up to 240mm tall. **Leaves** 2, flat on or slightly above the ground, without hairs, up to 30mm broad. **Inflorescence** lax, more or less 1-sided, 4–24-flowered; stalk without bracts, minutely and thinly hairy. **Flowers** pure white or flushed purple at base. Sepals 1.5–3.5 × 0.5–1.5mm; petals 3–6.5 × 0.5–1.5mm, fused to the lip at the base; lip 5-lobed, lobes more or less linear, 3.5–9 × 0.5–1.5mm; spur coiled, 1.5–5.5mm long.

Distinguishable from *H. orthoceras* by the coiled spur and rectangular rather than triangular lobes in the centre of the lip.

A S O N D J F M A M J J

Common in damp places and on mossy boulders, usually in a sheltered position under bushes and trees, from near sea level to 1,300m. Flowers between June and October.

Holothrix pilosa

Joubertina 15.01.2011

Joubertina 14.01.2011

Uniondale 20.01.2013

Joubertina 14.01.2011

Gansbaai 24.10.2010

Fairly robust terrestrial up to 430mm tall. **Leaf** 1, upper surface without hairs, margins and lower surface hairy, up to 140mm broad, usually withered at flowering time. **Inflorescence** lax to dense, more or less 1-sided, many-flowered; stalk without bracts, densely covered with long, soft hairs at right angles. **Flowers** resupinate or not, creamy white with a green central stripe. Sepals 2.5–5 × 1–2.5mm; petals 5.5–10.1 × 0.5–1mm; lip divided into 3–7 short lobes, 5.5–9 × 4.5–9mm; spur straight or curved, 1.5–2mm long.

A S O N D J F M A M J J

Localized in semi-arid, stony places in fynbos and renosterveld, from 35–1,000m. Flowers between the end of October and March.

Holothrix randii

Johannesburg 11.09.2012

Johannesburg 22.09.2011

Johannesburg 17.09.2012

Johannesburg 17.09.2012

Slender terrestrial up to 350mm tall. **Leaves** 2, pressed flat on the ground, without hairs, up to 40mm broad, withered at flowering time. **Inflorescence** lax, more or less 1-sided, 14–30-flowered; stalk with 3–6 bracts, minutely hairy. **Flowers** white, sepals green. Sepals 2.5–4.5 × 1.5–2mm; petals divided into 7–11 thread-like lobes, 6–13.5 × 2.8–5mm; lip divided into 8–15 hair-like lobes, 10–13.5 × 2.5–5.8mm; spur curved to nearly coiled, 3.5–8mm long.

A S O N D J F M A M J J

Very rare on grassy slopes and rock ledges, from 1,200–1,600m. Flowers between September and October.

Holothrix schlechteriana

Peddie 11.11.2011

Port Alfred 21.11.2011

Port Alfred 30.10.2010

Joubertina 19.01.2013

Slender to moderately robust terrestrial up to 400mm tall. **Leaves** 2, pressed flat on the ground, without hairs, up to 100mm broad. **Inflorescence** lax, more or less 1-sided, up to 50-flowered; stalk with 4–7 bracts, moderately to densely covered with short, fine, velvety hairs. **Flowers** green to pale greenish-cream to ochre. Sepals 2–6 × 0.5–2.5mm; petals divided into 4–9 lobes, 2.5–10 × 1–2.5mm; lip divided into 5–11 lobes, 6–7mm long; spur curved, 2.5–5.5mm long.

Distinguishable from *H. grandiflora* by the longer spur in relation to the lip, the hairy inflorescence stalk with fewer bracts and the earlier flowering time.

| A | S | O | N | D | J | F | M | A | M | J | J |

Localized among rocks and shrubs, from near sea level to 1,500m. Flowers between October and February.

Holothrix scopularia

Verloren Vallei 03.01.2011

Naudesnek Pass 11.12.2008

Verloren Vallei 06.01.2011

Naudesnek Pass 15.12.2013

Verloren Vallei 08.01.2013

Slender terrestrial up to 330mm tall. **Leaves** 2, pressed flat on the ground, with scattered hairs near the margin, up to 52mm broad, often withered at flowering time. **Inflorescence** fairly dense, 30–80-flowered, 1-sided; stalk without bracts, with long, slightly deflexed hairs. **Flowers** whitish or rich cream, occasionally suffused with pink. Sepals without hairs or with a few long hairs, 1.5–4 × 1–2mm; petals 3-lobed, 3–11 × 0.5–2.5mm; lip divided into 5–12 lobes, 3–13 × 1.5–4.5mm; spur curved, 1–4.5mm long.

A S O N D J F M A M J J

Localized on grassy slopes and rocky outcrops, from 1,200–2,700m. Flowers between September and January.

Holothrix secunda

Nieuwoudtville 07.10.2007

Nieuwoudtville 18.09.2011

Worcester 25.09.2012

Worcester 25.09.2012

Slender terrestrial up to 300mm tall. **Leaves** 2, pressed flat on the ground, without hairs. **Inflorescence** lax, more or less 1-sided, 12–27-flowered; stalk without bracts, slightly to densely hairy with short fine hairs. **Flowers** cream to yellow-green. Sepals without hairs, 1.5–3.5 × 1–2mm; petals attached to lip at their bases, 2.5–7 × 0.5–2mm; lip divided into 5 almost equal lobes, 3.5–8.5 × 1.5–4mm; spur slightly curved, 1.5–4.5mm long.

The hairless leaves and petals attached to the base of the lip distinguish this species from others with a 5-lobed lip.

A S O N D J F M A M J J

Common in dry areas, in the shade of bushes in stony soil, in rock crevices and on rock ledges, from 100–1,350m. Flowers between June and October.

Holothrix thodei

Elliot 25.01.2010

Sabie 24.01.2014

Sabie 24.01.2014

Sabie 24.01.2014

Slender terrestrial up to 250mm tall. **Leaves** not known. **Inflorescence** lax, more or less 1-sided, up to 40-flowered; stalk without bracts, densely covered with stout, reflexed hairs. **Flowers** yellowish-green. Sepals usually fused, with dense short hairs, 1.5–3.5 × 0.5–1mm; petals 3.5–5 × 0.5–1.5mm; lip 3-lobed, outer lobes one fifth to half as long as midlobe, 2.5–4.5 × 1.5–3.5mm; spur slightly curved, 0.5–2.5mm long.

Distinguishable from *H. incurva* by the absence of leaves at flowering time.

Occasional on basalt-derived soils in rocky grassland or in rock crevices, from 1,500–2,400m. Flowers between January and February.

Holothrix villosa

Slender terrestrial up to 310mm tall. **Leaves** 2, pressed flat on the ground, with scattered long hairs on upper surface, up to 95mm broad. **Inflorescence** lax to dense, more or less 1-sided, up to 100-flowered; stalk without bracts, densely to sparsely hairy with long hairs at right angles. **Flowers** cream to yellow-green. Sepals 1–3.5 × 0.5–1.5mm; petals entire, 1.5–7 × 0.5–1.5mm; lip with 3 almost equal lobes, 1.5–6.5 × 0.5–3.5mm; spur curved, 2–5.5mm long.

Holothrix villosa var. *condensata*

Syn: *Holothrix condensata*

Grabouw 14.11.2009

Cape Peninsula 06.12.2007

George 13.12.2010

Var. *condensata* is distinguished as follows: stalk stout, 85–240mm tall; inflorescence dense, up to 30-flowered; sepals often more than half the length of the petals, 1.5–3.5mm long; petals 3.5–7mm long; lip 3.5–6.5 × 1.5–3.5mm.

Var. *condensata* is most common in the south-western Cape Floral Region, with some records as far as George and Humansdorp.

Occasional (far less common than var. *villosa*) in peaty soil on rock seepages, mostly from 400–1,000m. Flowers between October and January, stimulated by fire.

A S O N D J F M A M J J

Holothrix villosa var. villosa

Piketberg 06.10.2007

Greyton 29.09.2010

Nieuwoudtville 10.09.2006

Tsitsikamma 05.11.2007

Avontuur 20.10.2013

Var. *villosa* is distinguished as follows: stalk slender, up to 365mm tall; inflorescence lax, up to 100-flowered; sepals 1–2.5mm long; petals 1.5–4.5mm long; lip 1.5–4 × 0.5–3mm.

Var. *villosa* is common in the Cape Floral Region and has also been recorded from Zimbabwe and Réunion.

Distinguishable from *H. cernua* by the softer, horizontal hairs on the inflorescence stalk, and from *H. exilis* by the subequal lip lobes.

A S O N D J F M A M J J

Very common in rock crevices and on hillsides in fynbos, from near sea level to 2,000m. Flowers between August and December.

Huttonaea fimbriata

Royal Natal National Park 18.01.2008

Royal Natal National Park 18.01.2008

Giant's Castle 27.01.2010

Slender terrestrial 170–320mm tall. **Leaves** 2, widely spaced; lower with leaf stalk 25–80mm long, very broadly egg-shaped with a notch at the base, 60–90 × 40–70mm; upper smaller and narrower. **Inflorescence** lax, 4–20-flowered; bracts as long as the ovaries. **Flowers** 10mm in diameter, white. Sepals unequal, margins notched with regular, rounded teeth, median reflexed, 4mm long, laterals spreading, 4–4.5 × 3.5mm; petals with claw 1mm long, blade fan-shaped, showy and shortly-fringed, 4–5mm long; lip fan-shaped, showy and shortly-fringed, 6 × 5mm.

A S O N D J F M A M J J

Locally common in cool forests, from 1,400–2,000m. Pollinated by oil-collecting bees. Flowers between January and March.

Huttonaea grandiflora

Naudesnek Pass 25.02.2010

Naudesnek Pass 05.02.2008

Sani Pass 23.02.2010

Naudesnek Pass 25.01.2009

Slender terrestrial 90–250mm tall. **Leaves** 2, widely spaced, sessile; lower broadly elliptic with a notch at the base, 20–60 × 15–40mm; upper much reduced, often present only as a sheath. **Inflorescence** lax, 1–5 flowered; bracts somewhat shorter than the ovaries. **Flowers** 20–30mm in diameter, white, petals with some mauve markings. Sepals unequal, margins toothed, median strap-shaped, 8mm long, laterals somewhat square, 7–8mm long; petals erect with blades curved forward, claws 7–8mm long, largely united with median sepal, blades fan-shaped, shallowly pouch-shaped with a callus at the base, finely and deeply fringed, up to 12mm long; lip elliptic, margins deeply fringed.

A S O N D J F M A M J J

Distinguishable from *H. oreophila* by larger flowers, longer petal fringes and completely fused petal claws.

Frequent in damp alpine grassland and in rock crevices, from 2,300–2,800m. Pollinated by oil-collecting bees. Hybridizes rarely with *H. oreophila*. Flowers between the end of January and March.

Huttonaea oreophila

Witsieshoek 20.02.2010

Witsieshoek 28.02.2013

Witsieshoek 28.02.2013

Witsieshoek 28.02.2013

Slender terrestrial 150–250mm tall. **Leaves** 2, widely spaced, sessile; lower broadly elliptic, 30–50 × 25–35mm; upper reduced or only present as a sheath. **Inflorescence** lax, 2–4-flowered; bracts shorter than ovary. **Flowers** 10–18mm in diameter, white, petals and lip with purple markings. Sepals unequal, margins finely toothed, median strap-shaped, united with the claws of the petals, laterals somewhat square, 6mm long; petals erect with blades often curved forward, claws 4–5mm long and largely fused, blades fan-shaped, concave with a callus near the base, margins finely fringed; lip elliptic, finely and irregularly fringed, 7 × 12mm.

Distinguishable from *H. grandiflora* by smaller flowers, shorter petal fringes and petal claws that are fused only at the base.

A S O N D J **F M** A M J J

Very local in high-altitude damp grassland, from 1,300–2,200m. Pollinated by oil-collecting bees. Hybridizes rarely with *H. grandiflora*. Flowers between February and March.

Huttonaea pulchra

Giant's Castle 26.02.2008

Giant's Castle 04.02.2007

Royal Natal National Park 22.02.2010

Slender to fairly robust terrestrial 150–400mm tall. **Leaves** 2, widely spaced, sessile; lower elliptic with a slight notch at the base, 60–120 × 45–70mm; upper smaller. **Inflorescence** lax, 6–20-flowered; bracts somewhat shorter than ovary. **Flowers** 15mm in diameter, sepals lime-green, petals pale mauve. Sepals similar, reflexed, margins entire, median 5–6 × 2.5–3mm, laterals spear-shaped, 7 × 4mm; petals with cylindrical claws of 5–7mm long and blade of 6–7mm in diameter, with margins fringed; lip egg-shaped, margins fringed, 4–5 × 7–10mm.

A	S	O	N	D	J	F	M	A	M	J	J

Rare in cool, dark forests from 1,000–1,800m. Pollinated by oil-collecting bees. Flowers between January and March.

Huttonaea woodii

Giant's Castle 27.01.2010

Giant's Castle 27.01.2010

Giant's Castle 27.01.2010

Giant's Castle 27.01.2010

Slender terrestrial up to 300mm tall. **Leaves** 2, widely spaced, sheathing at the base, narrowly elliptic, 40–60 × 20–25mm, upper one smaller. **Inflorescence** rather dense, 15–25-flowered; bracts taller than ovaries. **Flowers** 12mm in diameter, lip and sepals lime-green, petals speckled and lined with purple. Sepals unequal, reflexed, margins entire, median strap-shaped, basal half united with petal claws, 5mm long, laterals unequally egg-shaped, 6 × 4mm; petals with claw and blade equal in length, margins fringed; lip claw 3 × 4mm, blade egg-shaped, margins irregularly fringed.

| A | S | O | N | D | J | F | M | A | M | J | J |

Very rare in damp to marshy grassland, from 1,200–2,100m. Pollinated by oil-collecting bees. Flowers between January and February.

Jumellea walleri

Syn: *Jumellea filicornoides*

Tzaneen 02.03.2012

Tzaneen 11.03.2013

Tzaneen 22.02.2014

Tzaneen 02.03.2012

Slender to robust epiphyte, erect or straggling; stems up to 300mm long, 5mm in diameter, often branched at the base. **Leaves** linear to strap-shaped, keeled at the base, flat apically, unequally bilobed, 60–90 × 6–10mm. **Inflorescences** several, lateral, single-flowered. **Flower** 30mm in diameter, white becoming orange with age, ovary often with a sharp kink directly below the flower. Sepals 15–22 × 3–4mm; median sepal erect to reflexed; lateral sepals spreading; petals spreading, 12–16 × 2–3mm; lip rhomboidal, 20 × 6–7mm; spur 38–36mm long.

A S O N D J F M A M J J

Localized but often abundant in subtropical and submontane forests, from 500–1,000m. Flowers between February and March.

Liparis bowkeri

Graskop 13.12.2012

Graskop 23.01.2012

Royal Natal National Park 18.01.2008

Royal Natal National Park 04.01.2009

Slender terrestrial, or sometimes an epiphyte, 100–200mm tall; pseudobulbs adjacent, often well developed only in previous year's shoot. **Leaves** 2 or 3, egg-shaped to elliptic, with 5–7 prominent veins below, thin-textured, 45–120 × 25–70mm. **Inflorescence** lax, 4–12-flowered. **Flowers** green or yellowish-green, fading to orange. Sepals unequal; median sepal erect, wiry, 10–12 × 1.5–2mm; lateral sepals spreading, oblong, 8–11 × 4–5mm, free from each other along inner margin; petals hanging to spreading, linear, wiry, 8–12 × 1mm; lip very broadly egg-shaped, rounded, concave with a small conical callus at the base, 7–8 × 7–8mm. This species shows considerable variation over its distribution range.

A S O N D J F M A M J J

Distinguishable from *L. remota* by the adjacent pseudobulbs, 5–7 prominent veins on the underside of the leaves, lateral sepals free to the base and the larger flowers.

Locally common on forest floors, from near sea level to 1,900m. Flowers between November and March.

Liparis capensis

Cape Peninsula 04.06.2011

Cape Peninsula 03.06.2011

Cape Peninsula 03.06.2011

Cape Peninsula 03.06.2011

Slender terrestrial up to 100mm tall; pseudobulbs largely buried in the soil, adjacent, ovoid. **Leaves** 2 or 3, spreading near the soil surface, broadly egg-shaped, flat, firm-textured, 20–47 × 15–45mm. **Inflorescence** lax, 7–20-flowered. **Flowers** green or yellowish-green. Sepals unequal; median sepal reflexed, parallel to the ovary, 4–5 × 1.5mm, margins rolled; lateral sepals oblong, facing each other, 4 × 2mm; petals linear, margins rolled, spreading downward, 4 × 1mm; lip oblong, somewhat widened in the middle forming obscure side lobes, sharply notched, deeply concave with a small conical callus near the base, 4 × 2mm.

A	S	O	N	D	J	F	M	A	M	J	J

Occasional in full sun on sandy soil, in fynbos, from sea level to 600m. Flowers mostly between April and July.

Liparis remota

Oribi Gorge Nature Reserve 11.01.2011

Oribi Gorge Nature Reserve 11.01.2011

Nature's Valley 16.02.2012

Slender terrestrial, or sometimes an epiphyte, up to 200mm tall; pseudobulbs 20–50mm apart along a creeping stem. **Leaves** 3, ascending, elliptic, with 3 prominent veins below, thin-textured, 25–100 × 15–45mm. **Inflorescence** lax, 3–6-flowered. **Flowers** green. Sepals unequal; median sepal erect, strap-shaped, 6–9 × 2mm; lateral sepals projecting away, broadly elliptic, 5–6 × 3–4mm, largely united; petals deflexed, linear, margins rolled, 6–8 × 1mm; lip spatula-shaped; limb ascending, 2 × 2mm; apical lobe wider.

Distinguishable from *L. bowkeri* by the well-spaced pseudobulbs, 3 prominent veins on the underside of the leaves, lateral sepals free only in the apical third and the smaller flowers.

A S O N D J F M A M J J

Locally abundant in forests, from sea level to 1,000m. Flowers between November and March.

Margelliantha caffra

Syn: *Diaphananthe caffra; Mystacidium caffrum*

Nkandla 10.11.2012

Nkandla 17.12.2012

Nkandla 17.12.2012

Nkandla 21.12.2013

Miniature epiphyte with stems up to 20mm long. **Leaves** few, occasionally absent, linear to strap-shaped, unequally bilobed, 25–65 × 5–8mm. **Inflorescences** lateral, 1 to several, hanging from below the leaves, 20–45mm long, densely 5–15-flowered. **Flowers** white, anther cap green. Sepals spreading, 4.5–7 × 1.7–3mm; petals 5–6 × 1.8–3.3mm; lip 5–6 × 2.5–5mm, with small tooth at entrance to the spur; spur 12–17mm long, inflated near the tip.

A S O N D J F M A M J J

Occasional in montane forest, from 1,200–1,800m. Flowers between October and January.

Microcoelia aphylla

Syn: *Solenangis aphylla*

KZN north coast 03.10.2012

KZN north coast 29.07.2013

KZN north coast 29.07.2013

KZN north coast 29.07.2013

KZN north coast 29.07.2013

Slender leafless epiphyte with stems up to 400mm long, 2mm in diameter, climbing with hook-like roots. **Inflorescences** lateral, many, all along the stem, 10–25mm long, 8–16-flowered. **Flowers** white, sepals tipped with rust-red. Sepals 3 × 1.5–1.8mm wide; petals linear, up to 3mm long; lip obscurely 3-lobed, 3 × 2mm; spur strongly curved, 4–5mm long.

A S O N D J F M A M J J

Rare in west-facing coastal dune forest, from near sea level to 100m. Flowers between July and August

Microcoelia exilis

KZN north coast 08.05.2013

KZN north coast 06.02.2012

KZN north coast 13.03.2014

KZN north coast 13.03.2014

KZN north coast 13.03.2014

Slender leafless epiphyte with stems up to 40mm long, 3mm in diameter; roots numerous, branching, twisted, tangled, up to 350 × 1–2mm, not tightly attached to the host. **Inflorescences** lateral, several, 50–150mm long, up to 50-flowered. **Flowers** white, 1.5mm in diameter. Sepals and petals similar, 1–1.5 × 0.4–0.8mm; lip rounded to almost rounded, 0.6–1.1 × 0.4–0.8mm; spur spherical, 0.6–1mm in diameter.

A S O N D J F M A M J J

Localized but often common in riverine, coastal dune and sand forest, from near sea level to 600m. Flowers between February and April.

Microcoelia obovata

KZN north coast 24.10.2012

KZN north coast 24.10.2012

KZN north coast 25.10.2013

KZN north coast 25.10.2013

Slender leafless epiphyte with stems up to 30mm long; roots undulating, smooth, mostly well attached to the host. **Inflorescences** lateral, up to 10 simultaneously, up to 90mm long, 20-flowered. **Flowers** white with a conspicuous yellow-brown column and anther cap, up to 14mm long. Sepals and petals similar, egg-shaped to oblong, up to 4–7 × 1.3–2mm; lip broadly egg-shaped, with a thickening on each side of spur entrance, 5–7.5 × 3–5mm; spur slender, curved, 3.3–5.6mm long.

A S O N D J F M A M J J

Very localized in woodland and riverine forest, from sea level to 250m. Flowers around October.

Mystacidium aliceae

Nkwenkwe 14.02.2004

Mandini 31.01.2014

Mandini 18.01.2013

Mandini 31.01.2014

Mandini 31.01.2014

Miniature twig epiphyte with stems 5–10mm long. **Leaves** linear to spear-shaped, obscurely unequally bilobed, 20–60 × 4–9mm. **Inflorescences** lateral, several, short, not arching, 15mm long, densely 3–5-flowered. **Flowers** pale green, translucent, 10mm in diameter. Sepals and petals similar, reflexed, 4–5 × 2mm; lip 3-lobed; lip midlobe sharply deflexed, 2.5–3 × 2mm; spur tapering, 8–11mm long.

Distinguishable from *M. flanaganii* and *M. pusillum* by the short and crowded inflorescence, and the spur shorter than 11mm.

A S O N D J F M A M J J

Local and rather rare in coastal forest, from near sea level to 500m. Flowers between November and March.

Mystacidium brayboniae

Soutpansberg 23.11.2012

Soutpansberg 23.11.2012

Soutpansberg 11.12.2012

Slender epiphyte with stems 5–20mm long. **Leaves** few or absent at flowering time, strap-shaped, obscurely unequally bilobed, 20–50 × 5–10mm. **Inflorescences** lateral, 1 to several, 20–40mm long, densely 2–10-flowered. **Flowers** 10–15mm in diameter, white, anther cap greenish-yellow. Sepals similar; median sepal narrowly oblong, 8–8.5 × 4mm; lateral sepals narrowly elliptic to spear-shaped, 10 × 4mm; petals egg-shaped, 8–8.5 × 4mm; lip 3-lobed; lip midlobe sharply deflexed; spur tapering, 18–25mm long.

| A | S | O | N | D | J | F | M | A | M | J | J |

Known only from forests in the Soutpansberg, from 800–1,500m. Flowers between October and December.

Mystacidium capense

Sedgefield 05.12.2008

Lydenburg 16.11.2012

Sedgefield 26.11.2013

Sedgefield 26.11.2013

Slender epiphyte with stems 10–25mm long. **Leaves** strap-shaped, unequally bilobed, 40–120 × 10–22mm. **Inflorescences**, lateral, 1 to several, 70–180mm long, laxly 7–13-flowered. **Flowers** white, 25–30mm in diameter. Sepals and petals similar, reflexed, linear to spear-shaped, 9–17 × 3–4mm; lip 3-lobed, 10–15 × 3–4mm; spur tapering, 35–60mm long.

Distinguishable from *M. venosum* by the longer leaves and spur, and particularly the flowering time.

A S O N D J F M A M J J

Frequent in montane and lowland forests, also in drier localities such as *Acacia* thorn bush, sometimes on succulent *Euphorbia* species, from near sea level to 1,300m. Pollinated by hawkmoths. Flowers between September and January.

Mystacidium flanaganii

Graskop 08.01.2014

Lydenburg 09.01.2011

Graskop 05.01.2014

Graskop 08.01.2014

Miniature twig epiphyte with stems up to 10mm long. **Leaves** 3–5, strap-shaped, very obscurely unequally bilobed, 12–35 × 3–4.5mm. **Inflorescences** lateral, 1–5, 20–45mm long, laxly 3–10-flowered. **Flowers** pale green, 5–7mm in diameter. Sepals unequal; median sepal strap-shaped, 2–3.5 × 0.8–1.2mm; lateral sepals narrowly rhomboid to spear-shaped, 2.6–5 × 1–1.5mm; petals triangular, 1.6–3 × 0.8–1.5mm; lip 3-lobed; lip midlobe sharply deflexed, 1.5–2 × 0.8–1mm; lip side lobes poorly developed; spur tapering, 18–25mm long.

Difficult to separate from *M. pusillum* on morphological characters and distinguishable only by flowering time. Distinguishable from *M. aliceae* by the longer inflorescence with evenly spaced flowers and the spur longer than 18mm, and from *M. gracile* by the flowering time.

A	S	O	N	D	J	F	M	A	M	J	J

Locally abundant in temperate and subtropical forests, from near sea level to 1,800m. Flowers between December and January.

Mystacidium gracile

Mariepskop 05.08.2013

Mariepskop 05.08.2013

Lydenburg 19.09.2010

Lydenburg 13.08.2012

Miniature epiphyte with stems 10–15mm long. **Leaves** rarely produced, 10–15 × 3mm; roots numerous, unbranched, mostly attached to the host. **Inflorescences** lateral, 1–4, 30–50mm long, laxly 10-flowered. **Flowers** pale green, 8mm in diameter. Sepals and petals similar, apices more or less fleshy, 3–5 × 0.7–1.5mm; lip 3-lobed; lip midlobe sharply deflexed; spur tapering, 17–25mm long.

Similar to *M. flanaganii* or *M. venosum* if it produces leaves. Distinguishable from *M. flanaganii* by the much later flowering season, whereas the flowers are about half the size of those of *M. venosum*. Distinguishable from *M. capense* by the fleshy tepal apices.

A S O N D J F M A M J J

Widespread and often common on tree trunks, in montane or temperate forests, from 1,300–1,800m. Flowers between June and October.

Mystacidium pusillum

Mandini 10.06.2013

Mandini 10.06.2013

Mandini 06.06.2013

Mandini 19.06.2012

Miniature twig epiphyte with stems shorter than 10mm. **Leaves** 1–3, strap-shaped, very obscurely unequally bilobed, 15–40 × 4–6mm. **Inflorescences** lateral, 1 to several, 15–35mm long, 4–7-flowered. **Flowers** pale green, 5mm in diameter. Sepals unequal; median sepal 2.6–5 × 1mm; lateral sepals spreading, spear-shaped to rhomboid, 3–5 × 0.7–2mm; petals egg-shaped, 2–3 × 1–2mm; lip midlobe triangular, 1mm long; lip side lobes absent; spur tapering, 12–18mm long.

Difficult to separate from *M. flanaganii* on morphological characters and distinguishable only by flowering time.

| A | S | O | N | D | J | F | M | A | M | J | J |

Occasional and rather local in temperate or subtropical forests, from near sea level to 1,200m. Flowers between June and August.

Mystacidium venosum

Eshowe 17.06.2012

Eshowe 15.07.2012

Graskop 10.05.2013

Graskop 11.04 .2013

Miniature epiphyte with stems 10–15mm long. **Leaves** strap-shaped to elliptic, unequally bilobed, 15–45 × 6–10mm, often poorly developed, occasionally absent. **Inflorescences** lateral, 1 to several, 20–50mm long, 4–10-flowered. **Flowers** white, 15mm in diameter. Sepals reflexed to spreading, 7–9 × 2–3mm; petals reflexed, slightly shorter and narrower; lip 3-lobed; lip midlobe sharply deflexed; spur tapering, 20–45mm long.

Distinguishable from *M. capense* by the shorter leaves and spur, and particularly the flowering time.

| A | S | O | N | D | J | F | M | A | M | J | J |

Occasional in montane and coastal forests, from near sea level to 1,700m. Pollinated by hawkmoths. Flowers between April and July.

Neobolusia tysonii

Port St Johns 12.12.2011

Cathedral Peak 19.01.2008

Sabie 24.01.2013

Slender terrestrial up to 400mm tall. **Leaves** 2, basal, spreading, narrowly elliptic to oblong, up to 155 × 20mm; a few small, sheathing leaves higher up. **Inflorescence** lax, up to 150mm long, 2–12-flowered. **Flowers** with sepals brownish-green, petals off-white, lip white with pink marking at the base; 15 × 6mm. Sepals similar, 9 × 3mm; petals 7 × 3mm, partially enclosing the column; lip spatula-shaped, up to 12 × 8mm, margin finely wavy to finely toothed.

A S O N D J F M A M J J

Common in moist or marshy montane grassland, from 350–2,350m. Flowers between December and February.

Nervilia bicarinata

Syn: *Nervilia umbrosa*

Ugutugulu (Swaziland) 31.10.2012

Ugutugulu (Swaziland) 16.11.2012

Ugutugulu (Swaziland) 31.10 .2012

Ugutugulu (Swaziland) 14.12 .2012

Fairly robust terrestrial 170–750mm tall. **Leaf** 1, developing during or after flowering, held horizontally and well above the ground, circular to heart-shaped, 10–30-veined, pleated, hairless, up to 225 × 265mm; leaf stalk 50–260mm long. **Inflorescence** lax, 2–12-flowered. **Flowers** with sepals and petals greenish, lip greenish-white with purple or green veins. Sepals and petals similar, strap- to spear-shaped, 17–31 × 0.9–4mm; lip egg-shaped, 20–31 × 17–25mm, obscurely 3-lobed, with 2 fleshy ridges; lip midlobe triangular; lip side lobes short, erect, enclosing the column.

A S O N D J F M A M J J

Rare and sporadic in riverine and coastal swamp forest or between granite rocks under *Acacia ataxacantha*, from near sea level to 1,000m. Flowers between late October and November.

Nervilia crociformis

Alkmaar 23.11.2012

Alkmaar 23.11.2012

Alkmaar 23.11.2012

Alkmaar 15.12.2012

Very slender terrestrial up to 100mm tall. **Leaf** 1, developing during or after flowering, prostrate, held horizontally and close to the ground, kidney-shaped to circular, with 11 clearly visible veins, usually hairy above, 15–90 × 35–140mm. **Inflorescence** single-flowered, lengthening when fruiting. **Flower** more or less erect, sepals and petals brownish-green, lip white with yellow centre and often lilac or purple markings near the tip. Sepals and petals spreading, similar, linear to strap-shaped, 12–19 × 3.5mm; lip 12–18 × 9–11mm, 3-lobed; lip blade hairy, with 3 raised and thickened ridges.

A S O N D J F M A M J J

Rare and sporadic in woodland, evergreen forest and grassland, from 600–1,200m. Flowers between October and December.

Nervilia kotschyi

Pretoria 14.11.1982

Pretoria 14.11.1982

Pretoria 14.11.1982

Barberton 21.12.2013

Slender terrestrial 80–420mm tall. **Leaf** 1, developing after flowering, erect, elliptic to spear-shaped, base tapering or cut off squarely, 6–10-veined, pleated, hairless, 60–130 × 25–40mm; leaf stalk 30–60mm long. **Inflorescence** lax, 2–5-flowered. **Flowers** yellow to greenish, lip whitish-green with purple venation. Sepals and petals similar, strap- to spear-shaped, 12–26 × 3–3.5mm; lip elliptic, 10–19 × 7–12mm, 3-lobed; lip blade with 2 fleshy ridges.

Var. *purpurata* (syn: *Nervilia purpurata*) is widespread in Africa and the only variety occurring in South Africa.

A S O **N D** J F M A M J J

Very rare in grassland and savanna, from 1,200–1,500m. Flowers in November.

Nervilia lilacea

Lydenburg 12.11.2004

Lydenburg 12.11.2004

Lydenburg 12.11.2004

Very slender terrestrial 25–100mm tall. **Leaf** 1, developing after flowering, held horizontally and close to the ground, heart-shaped, hairless, 14–55 × 12–45mm; leaf stalk 10–60mm long. **Inflorescence** single-flowered, lengthening to 60–200mm when fruiting. **Flower** small, often self-pollinating within the unopened flower, borne horizontally, lasting only a few hours, sepals and petals reddish-green with maroon markings, lip white to faint pink with darker markings; 22 × 18mm when open. Sepals and petals spreading, similar, strap-shaped, 9.5–14 × 2mm; lip oblong, 3-lobed, 9–13 × 5–6mm, with a narrow waist.

A	S	O	N	D	J	F	M	A	M	J	J

Very rare in woodland and riverine forest, from 1,150–1,900m. Flowers between November and January.

Oeceoclades decaryana

Vryheid (cultivated plant) 14.01.2014

Vryheid (cultivated plant) 14.01.2014

KZN north coast 02.10.2012

Fairly slender terrestrial up to 600mm tall; pseudobulbs ovoid, 5-angled,
30 × 20mm. **Leaves** 2 or 3 per pseudobulb, sessile, linear, 180–220 × 12–20mm.
Inflorescence lax, 12–20-flowered. **Flowers** dark green with purple veins,
lip midlobe cream or pale green. Sepals spatula-shaped, 12–20 × 3–4mm;
petals 8 × 4mm; lip 3-lobed or obscurely 4-lobed, crest a transverse ridge;
lip midlobe divided into 2 lobules, lobes spreading, 8 × 6mm; spur cylindrical,
curved forward, 5mm long.

A S O N D J F M A M J J

Rare in leaf mould under bushes, among rocks and in sand forest, from near sea level to 200m. Flowers in December.

Oeceoclades lonchophylla

Mtunzini 29.01.2013

Mtunzini 29.01.2013

Richards Bay 15.01.2013

Richards Bay 23.10.2012

Richards Bay 01.12.2012

Slender terrestrial up to 280mm tall; pseudobulbs conical, 25 × 10mm. **Leaf** 1 on each pseudobulb, blade spear-shaped, margins wavy, 60–130 × 20–40mm; leaf stalk 75–150mm long. **Inflorescence** branched, lax, 20–30-flowered. **Flowers** with sepals lime-green to yellowish, petals similar but irregularly blotched reddish-purple, lip white with purple markings near the base. Sepals similar, 5.5–8 × 2–2.5mm; petals 5 × 3mm; lip 4-lobed; spur 2mm long.

| A | S | O | N | D | J | F | M | A | M | J | J |

Occasional on sandy soils, from near sea level to 500m. Flowers between January and April.

Oeceoclades maculata

Syn: *Oeceoclades mackenii*

Ngoye 14.02.2013

Ngoye 14.02.2013

Ngoye 15.07.2012

Ngoye 04.03.2013

Ngoye 20.03.2013

Slender terrestrial up to 300mm tall; pseudobulbs conical, 20–30 × 10–15mm.
Leaf 1 per pseudobulb, blade strap-shaped to elliptic, green speckled with
darker spots, 110–180 × 25–45mm; leaf stalk absent or up to 20mm long.
Inflorescence lax, simple or sometimes branched, 200–300mm tall,
up to 30-flowered. **Flowers** greenish-yellow, lip bright yellow, red-streaked.
Sepals spreading, 8–11 × 3.5–4mm; petals erect, 8–10 × 4–5mm; lip 4-lobed,
8–9 × 11mm, with crest of transverse ridges at entrance to spur; spur club-
shaped, rounded or sharply notched, 4–6mm long.

A S O N D J F M A M J J

Restricted and rare in leaf mould in dark shady forests, from near sea level to 300m. Flowers between January
and March.

Oeceoclades quadriloba

Siteki (Swaziland) 08.02.2013

Siteki (Swaziland) 24.02.2014

Siteki (Swaziland) 26.01.2013

Siteki (Swaziland) 08.02.2013

Slender terrestrial up to 220mm tall; pseudobulbs 12–20mm long, often under humus. **Leaves** 2, stalked, spear-shaped to linear, 40–70 × 2–4mm. **Inflorescence** 1 or 2, lax, branched, 5–18-flowered. **Flowers** with sepals and petals green, lip white to cream tinged yellow. Sepals and petals similar, 4mm long; lip 4-lobed, 6 × 5mm, bearing 2 triangular callus ridges in the mouth of the spur; spur club-shaped, 5–7mm long.

A S O N D J F M A M J J

Rare in rock crevices in forest, so far recorded only from Swaziland, from 100–500m. Flowers between late January and March.

Orthochilus aculeatus

Slender to fairly robust terrestrial 100–650mm tall. **Leaves** stiffly erect, pleated, partly to fully developed at flowering time, up to 600 × 17mm. **Inflorescence** short and dense, 3–27-flowered. **Flowers** dull ivory, white, greenish-white to greenish-pink, or dark reddish-purple. Sepals and petals scarcely spreading, similar; median sepal oblong to elliptic-oblong, 6–16mm long; lateral sepals slightly longer and broader; crests of lip consisting of 2 ridges in the basal half, becoming slender flattened papillae on the midlobe, spurless; column foot 0.6–5mm long, column 3–5mm long.

Orthochilus aculeatus subsp. aculeatus

Syn: *Eulophia aculeata* subsp. *aculeata*

Ngeli 07.11.2013

Sabie 18.11.2013

Storms River Mouth 03.12.2007

Sabie 18.11.2013

Sabie 18.11.2013

Subsp. *aculeatus* is distinguished as follows: its shorter stature, usually 100–200mm tall; leaves up to 11mm wide; flowers nodding; median sepal and petals 6–8mm long; column foot 0.6–2.1mm long.

Subsp. *aculeatus* is restricted to South Africa, Lesotho and Swaziland.

Distinguishable from *O. leontoglossus* and *O. vinosus* by the absence of a spur.

Common in fynbos in the coastal belt of the Cape Floral Region and in sour grassland in the summer-rainfall areas, from near sea level to 2,200m. Flowers between November and January.

| A | S | O | N | D | J | F | M | A | M | J | J |

Orthochilus aculeatus **subsp.** *huttonii* Syn: *Eulophia aculeata* subsp. *huttonii*; *Eulophia huttonii*

Naudesnek Pass 18.12.2010

Garden Castle 08.12.2012

Naudesnek Pass 18.12.2010

Dullstroom 01.11.2012

Subsp. *huttonii* is distinguished as follows: its taller stature, usually over 200mm; leaves up to 17mm wide; flowers held horizontally; median sepal 12–16mm long; petals 9.3–14.5mm long; column foot 2.3–5mm long.

Subsp. *huttonii* is restricted to South Africa.

Distinguishable from *O. leontoglossus* and *O. vinosus* by the absence of a spur.

| A | S | O | N | D | J | F | M | A | M | J | J |

Localized to fairly common in sour grassland, rarely in tall grassland and coastal thornveld, from 1,000–2,100m. Flowers between October and January.

Orthochilus adenoglossus

Syn: *Eulophia adenoglossa; Eulophia nigricans*

Cato Ridge 12.01.2014

Cato Ridge 15.01.2013

Cato Ridge 08.01.2012

Cato Ridge 08.01.2012

Slender to robust terrestrial 300–870mm tall. **Leaves** pleated, absent to partly developed at flowering time, up to 520 × 8mm. **Inflorescence** rather lax, 4–22-flowered. **Flowers** with sepals yellow-green to olive-green, petals paler, lip pale purple with main uncrested veins and distal crest papillae dark purple. Sepals and petals similar, partly spreading; median sepal 17–21mm long; lip 3-lobed, crests consisting of 2 broad, low ridges in basal two thirds, becoming a few rows of irregular papillae on the midlobe; spur cylindrical to slightly conical, 2.5–5mm long; column 9–12mm long, column foot 3–6mm long.

A S O N D J F M A M J J

Very rare in grassland, from 600–1,500m. Flowers between December and January.

Orthochilus chloranthus

Syn: *Eulophia chlorantha*

Josefsdal 27.10.2012

Josefsdal 27.10.2012

Josefsdal 27.10.2012

Mbabane (Swaziland) 05.11.2011

Slender terrestrial 90–350mm tall. **Leaves** stiffly erect, partly to fully developed at flowering time, up to 370 × 5mm. **Inflorescence** somewhat lax, 5–18-flowered. **Flowers** with sepals green, petals and lip pale green, lip side lobes, basal petal margins and column whitish. Sepals similar, partly spreading, 8–11mm long; petals slightly shorter; lip 3-lobed, crests consisting of 2 finely hairy blades in the basal two thirds, becoming irregular papillae on the rounded midlobe; spur stout, 1.2–1.8mm long; column 3–4mm long, column foot 1–1.5mm long.

A S O N D J F M A M J J

Rare on exposed grassy slopes or in sheltered grassy areas in bushveld (where it assumes a taller and more robust habit), from 1,200–1,600m. Flowers between September and December.

Orthochilus ensatus

Syn: *Eulophia ensata*

Warburton 09.01.2011

Lochiel 13.01.2013

Mtunzini 09.11.2012 (white form)

Warburton 21.12.2010

Slender terrestrial 300mm–1m tall. **Leaves** stiffly erect, pleated, partly to fully developed at flowering time, up to 900 × 15mm. **Inflorescence** short and very dense, 6–30-flowered. **Flowers** pale to bright yellow, rarely white, crests on lip a slightly deeper yellow. Sepals and petals almost bell-shaped to partly spreading; sepals 16–23mm long; petals slightly shorter and narrower; lip 3-lobed, crests consisting of 2 broad ridges in the basal half, becoming a dense mass of long papillae extending almost to the apex; spur 4–7mm long; column 6–8mm long, column foot 2mm long.

Distinguishable from *O. welwitschii* by the crest papillae extending almost to the apex, the smaller flowers, the slender column and the lip side lobes that are the same colour as the rest of the flower.

a | A S O N D J F M A M J J
b | A S O N D J F M A M J J

Rare to locally frequent in open grassland, grassy areas in coastal bushveld and moderate shade in pure bushveld, from near sea level to 1,800m. Pollinated by scarab beetles. Yellow form flowers between December and January (**a**); white form flowers between August and December (**b**).

Orthochilus foliosus

Syn: *Eulophia foliosa*

Dullstroom 02.12.2011

Maclear 17.12.2010

Port Edward 16.08.2012

Fairly stout terrestrial 150–450mm tall. **Leaves** pleated, rather stiffly erect, partly to fully developed at flowering time, up to 600 × 25mm. **Inflorescence** elongate and dense, 6–40-flowered. **Flowers** with sepals and petals dull lime-green to yellowish-green, distal part of lip tinged with dark purple, or rarely pale purple to white. Sepals and petals almost bell-shaped, 10–13mm long; petals slightly broader than sepals; lip 3-lobed, crests consisting of 2 or 3 broad ridges, becoming a few very short, stout papillae confined to basal part of midlobe; spur absent; column 3.5–5mm long, column foot 2–3mm long.

A S O N D J F M A M J J

Frequent in sour grassland, occasionally in grassy areas in coastal thornveld, from near sea level to 2,000m. Pollinated by click beetles. Flowers between August and January.

Orthochilus leontoglossus

Syn: *Eulophia leontoglossa*

Loteni 05.01.2009

Port Edward 24.07.2013

Graskop 01.01.2011

Peacevale 01.12.2013

Boston 09.12.2011

Slender to fairly stout terrestrial 60–280mm tall. **Leaves** pleated and stiffly erect, partly to fully developed at flowering time, up to 370 × 9mm. **Inflorescence** short and dense, axis curved to nodding, 7–35-flowered. **Flowers** yellow, white or white tinged with purple, papillae often differently coloured. Sepals and petals similar, slightly spreading near the tip, 10–14mm long; lip 3-lobed, crests consisting of 2–4 very low ridges in basal two thirds, becoming several rows of slender papillae; spur 3–5mm long; column 4–6mm long, column foot rudimentary.

A S O N D J F M A M J J

Distinguishable from *O. vinosus* by the later flowering time, uppermost bracts overtopping the inflorescence, and white or yellow-orange (rarely pale pink) flowers. Distinguishable from *O. aculeatus* subsp. *aculeatus* by the presence of a spur.

Rare to locally frequent in dry, moderately moist and occasionally marshy areas in grassland, from near sea level to 2,200m. Flowers between August and January.

Orthochilus litoralis

Stellenbosch 07.12.2009

Stellenbosch 07.12.2009

Stellenbosch 07.12.2009

Kleinmond 17.12.2013

Slender terrestrial 220–660mm tall. **Leaves** mostly absent, occasionally a single blade up to 20mm long. **Inflorescence** lax, 6–27-flowered. **Flowers** mostly resupinate, sepals yellowish-green, faintly tinged with olive, petals yellow, lip yellow, side lobes tinged with purple, crests bright yellow. Sepals and petals partly spreading; sepals 17–24mm; petals slightly shorter or longer; lip 3-lobed, crests consisting of 2 fleshy ridges in basal third, becoming several rows of stiffly erect papillae on the midlobe; spur 2.5–4.5mm long; column 10–13mm long, column foot nearly absent to 2mm long.

| A | S | O | N | D | J | F | M | A | M | J | J |

Rare in sandy soils in a narrow belt along the coast, mostly near sea level, but recorded to 850m. Flowers between November and January.

Orthochilus milnei

KZN south coast 06.12.2012   KZN south coast 11.12.2011   KZN south coast 30.11.2013

KZN south coast 06.12.2012

Slender terrestrial 180–520mm tall. **Leaves** very slender, partly to fully developed at flowering time, up to 390 × 2–3mm. **Inflorescence** rather dense, 5–19-flowered. **Flowers** dull to bright yellow, lip sometimes slightly paler. Sepals and petals partly curved; sepals 4–7mm long; petals nearly the same length as the median sepal; lip 3-lobed, crests consisting of 2–4 very low ridges on basal half, becoming several rows of slender papillae further on; spur 2–4mm long; column 1.5–3.5mm long, column foot absent.

A S O N D J F M A M J J

Rare and scattered in swampy soils, sometimes in free-standing water, from near sea level to 1,800m. Flowers between November and January.

Orthochilus odontoglossus

Syn: *Eulophia odontoglossa*

Dullstroom 02.01.2012

Eshowe 18.01.2013

Dullstroom 12.01.2011

Dullstroom 12.01.2010

Slender terrestrial 500–900mm tall. **Leaves** pleated and stiffly erect, partly to fully developed at flowering time, up to 1m × 9mm. **Inflorescence** somewhat dense, 10–30-flowered. **Flowers** pale to bright yellow, lip sometimes slightly darker, base of midlobe papillae and side lobes of lip sometimes reddish-brown. Sepals and petals spreading; sepals 8–12mm long; petals slightly shorter and narrower than the median sepal; lip 3-lobed, crests consisting of 2 blades in upper half, with numerous slender papillae; spur rudimentary, rarely reaching a length of 0.8mm; column 3.5–6.5mm long, column foot 2–4mm long.

A S O N D J F M A M J J

Locally frequent in coastal thornveld and at higher elevations in sour grassland, from near sea level to 2,200m. Flowers between October and February.

Grabouw 02.01.2010

Grabouw 02.12.2010

Swellendam 21.12.2012

Swellendam 21.12.2012

Slender terrestrial 100–400mm tall. **Leaves** erect and pleated, absent to partly developed at flowering time, up to 150 × 7mm. **Inflorescence** dense, 2–12-flowered, young flowers not fully resupinate. **Flowers** pale lemon, crests bright orange-yellow. Sepals and petals partly spreading; sepals 20–22mm long; petals slightly shorter and narrower; lip 3-lobed, crest consisting of a single, minutely hairy ridge for most of the lip length, becoming 2 ridges at the base, and ending in 2 or 3 lobes on the midlobe; spur absent; column 8–10mm long, column foot 2–3mm long.

| A | S | O | N | D | J | F | M | A | M | J | J |

Rare in fynbos in sandy soils derived from the Table Mountain Group, sometimes in marshy areas with a large admixture of humus, from 400–1,800m. Flowers between November and January, stimulated by fire.

Orthochilus vinosus

Dullstroom 08.10.2012

Dullstroom 05.11.2011

Dullstroom 07.10.2012

Dullstroom 08.10.2012

Slender terrestrial 120–210mm tall. **Leaves** pleated and stiffly erect, just emerging or partly developed at flowering time. **Inflorescence** short and dense, axis always erect, 9–18-flowered. **Flowers** pink to plum, paler on the inside, crest papillae yellow. Sepals and petals similar, only slightly spreading near the tip, 12–17 × 5–7mm; lip 3-lobed, crests consisting of 4 low ridges, becoming 5–8 rows of slender papillae; spur 4mm long; column 4–6mm long, column foot rudimentary.

Distinguishable from *O. leontoglossus* by the earlier flowering time, the uppermost bracts not reaching the inflorescence, and the purple flowers. Distinguishable from *O. aculeatus* subsp. *aculeatus* by the presence of a spur.

A S O N D J F M A M J J

Occasional to rare in high-altitude grassland, from 1,200–2,200m. Flowers between late September and early November.

Orthochilus welwitschii

Syn: *Eulophia welwitschii*

Giant's Castle 05.01.2009

Giant's Castle 16.12.2008

Commondale 04.12.2013

Elliot 10.01.2009

Slender to moderately robust terrestrial 260–880mm tall. **Leaves** stiffly erect, pleated, partly to fully developed at flowering time, up to 700 × 23mm. **Inflorescence** short and dense, 4–25-flowered. **Flowers** straw with dark reddish-purple on the lip side lobes, some of the crest papillae and the basal parts of the lip midlobe. Sepals and petals scarcely spreading, more or less bell-shaped; sepals 22–35mm long; petals slightly shorter than sepals and usually narrower; lip 3-lobed, crests consisting of 2 fleshy ridges on the basal third, becoming numerous thread-like papillae extending halfway along the midlobe; spur 2.7–6.8mm long; column 5–7mm long, column foot 1–2mm long.

A S O N D J F M A M J J

Fairly common in dry to marshy areas in grassland, from 200–1,800m. Pollinated by scarab beetles. Flowers between November and January.

Pachites appressus

Swellendam 29.12.2012

Swellendam 19.12.2012

Swellendam 19.12.2012

Swellendam 19.12.2012

Slender to fairly robust terrestrial up to 400mm tall. **Leaves** 5–12, linear, up to 100mm long. **Inflorescence** 4–32-flowered. **Flowers** non-resupinate, sepals and petals mauve-pink with a dark central stripe, lip greenish-yellow with dark mauve central stripe, margins and tip. Sepals 5–7mm long; petals similar to sepals; lip entire, slightly narrower than petals; column conspicuous, 5mm high, with 2 more or less erect projecting arms.

A S O N D J F M A M J J

Extremely rare on mountain slopes in fynbos, from 450–1,400m. Flowers between December and January, after fire.

Pachites bodkinii

Cape Peninsula 27.11.2007

Betty's Bay 29.11.2008

Porterville 08.12.2009

Cape Peninsula 27.11.2007

Slender terrestrial up to 200mm tall. **Leaves** 3–7, linear, up to 90mm long. **Inflorescence** rather dense, 3–20-flowered. **Flowers** non-resupinate, sepals and petals dull purplish-pink to pale green and iridescent white, margins purple to pale green, lip pale green near the base and dark purple near the tip. Sepals 12–15mm long, partly spreading to bending inward; petals similar to sepals; lip with small side lobes; column conspicuous, 6mm high.

A S O N D J F M A M J J

Rare, scattered or in small populations in fynbos, marshes, or on moist slopes, from 100–1,300m. Flowers between November and December, after fire.

Platycoryne mediocris

Nelspruit 18.01.2001

Slender terrestrial up to 300mm tall. **Leaves** 5–10, cauline, spear-shaped, 25–45 × 4–9mm. **Inflorescence** head-like, rather dense, 2–5-flowered. **Flowers** bright orange. Median sepal egg-shaped, forming a hood together with the petals, 7–11mm long; lateral sepals deflexed, 8–10mm long; petals entire, attached to median sepal; lip shortly 3-lobed near the base, 6.5–10 × 1–2mm; lip side lobes short, tooth-like; spur inflated apically, 9–17mm long.

| A | S | O | N | D | J | F | M | A | M | J | J |

Known from a single population near Nelspruit where it has been seen for a few years from 2001 onwards, growing in a seasonal seepage over granite at 800m. Flowers between January and February.

Platylepis glandulosa

Eshowe 15.02.2013

Eshowe 03.03.2012

Eshowe 07.03.2013

Eshowe 09.02.2012

Creeping terrestrial becoming erect, up to 500mm tall. **Leaves** present only in the lower half of the erect stem, stalked, with conspicuous veins, egg-shaped to elliptic, up to 40–90 × 20–50mm. **Inflorescence** congested, ovoid, conspicuously glandular-hairy, up to 100mm tall. **Flowers** green to brownish-green and white, lip white. Sepals 7–8 × 2–3mm; median sepal forming a hood with the petals; lateral sepals reflexed midway; lip fiddle-shaped, with 2 rounded sacs and a callus in each sac, reflexed apically, 6 × 4mm.

A S O N D J F M A M J J

In swamp forest and near streams in dense shade, from sea level up to 500m. Flowers between January and March.

Polystachya albescens

Tzaneen 12.05.2013

Sabie 29.10.2013

Tzaneen 28.10.2012

Sabie 07.02.2013

Fairly robust epiphyte 150–400mm tall; stems slender, not swollen into pseudobulbs. **Leaves** 2–4, thin, linear to strap-shaped, 120–220 × 8–16mm. **Inflorescence** shorter than the leaves, branched, densely many-flowered. **Flowers** white, cream to greenish-yellow, tinged with yellow and purple. Sepals unequal; median sepal narrowly triangular, 7.5–9 × 3–3.5mm; lateral sepals broadly egg-shaped, 8.5–10 × 5–8mm; petals strap- to spatula-shaped, 6–7 × 1.5–2.2mm; lip 3-lobed, hairy on top, 8 × 6–8mm, disc with a central knob-like callus; mentum conical, 6–7mm tall.

Subsp. *imbricata* is widespread in Africa and is the only subspecies occurring in South Africa and Swaziland.

Distinguishable from *P. transvaalensis* by the thin-textured, pointed leaves and much-branched inflorescence with more than 10 flowers.

A	S	O	N	D	J	F	M	A	M	J	J

Rare in montane forests, along streams, from 800–1,400m. Flowers between late October and February.

Polystachya concreta

Syn: *Polystachya mauritiana, Polystachya tessellata*

KZN north coast 14.01.2012

Mtunzini 06.02.2014

KZN north coast 18.01.2013

KZN north coast 29.01.2013

Slender to robust epiphyte or lithophyte up to 600mm tall; pseudobulbs narrowly conical, 20–40 × 10–15mm, enclosed by leaf sheaths. **Leaves** 3–5, narrowly elliptic, 80–210 × 25–50mm. **Inflorescence** dense, branched, usually much taller than the leaves, many-flowered; inflorescence stalk covered in papery sheaths. **Flowers** variable from creamy yellow to pale green, pink or dull red, lip white with midlobe pink. Sepals unequal; median sepal narrowly triangular, 3.5–4 × 2mm; lateral sepals obliquely egg-shaped, 4.5–5.5 × 4–4.5mm; petals strap- to spatula-shaped, 3–3.5 × 0.8–1mm; lip 3-lobed, 4.5–5 × 4–5mm, disc with a linear raised callus and powdery hairs, tapering towards the base; mentum conical, 3mm tall.

Distinguishable from *P. modesta* by an obvious callus on the disc of the lip.

| A | S | O | N | D | J | F | M | A | M | J | J |

Occasional in subtropical riverine forest, in full sun, partial or deep shade, from near sea level to 1,200m. Flowers between October and January.

428   *Polystachya*

Polystachya cultriformis

Kaapsehoop 12.03.2014

Kaapsehoop 12.03.2014

Ngeli (cultivated plant) 09.06.2013

Ngeli (cultivated plant) 23.07.2013

Slender to robust epiphyte or lithophyte up to 300mm tall; pseudobulbs adjacent, cylindrical to conical, 40–100 × 4–6mm. **Leaf** 1, narrowly spear-shaped, margins often wavy, 120–200 × 15–25mm. **Inflorescence** shorter or longer than the leaf, branched or rarely simple, 10–20-flowered. **Flowers** dull yellow. Sepals unequal; median sepal narrowly egg-shaped, 5–6 × 2–3mm; lateral sepals obliquely egg-shaped, 6–7 × 4mm; petals spatula-shaped, 4–5 × 1–1.5mm; lip 3-lobed, 5mm long, disc with a knob-like callus; mentum 3mm tall.

A S O N D J F M A M J J

Occasional in cool moist montane forests, from 1,000–1,800m. Flowers between October and January.

Polystachya fusiformis

Graskop 04.07.2013

Graskop 04.01.2014

Graskop 06.01.2014

Graskop 21.12.2013

Graskop 21.12.2013

Slender epiphyte up to 600mm tall; pseudobulbs slender, cylindrical, 80–160 × 2–8mm, superposed with the new pseudobulbs arising halfway along the old one. **Leaves** 3–5, narrowly elliptic, 70–100 × 15–30mm. **Inflorescence** dense, branched, up to 50-flowered; inflorescence stalk hairy. **Flowers** yellowish-green to reddish. Sepals unequal; median sepal narrowly triangular 3 × 1.8mm; lateral sepals obliquely triangular, 5 × 3mm; petals linear to strap-shaped, 3 × 0.6mm; lip 3-lobed, without a callus, 4.5 × 3.5mm; mentum conical, 2.6mm tall.

| A | S | O | N | D | J | F | M | A | M | J | J |

Localized but often common in cool montane forests, from 1,000–1,800m. Flowers between October and January.

430 *Polystachya*

Polystachya modesta

KZN north coast 03.10.2012

KZN north coast (cultivated plant) 28.01.2013

KZN north coast (cultivated plant) 28.01.2013

KZN north coast (cultivated plant) 28.01.2013

Slender epiphyte up to 400mm tall; pseudobulbs conical, 10–70 × 3–10mm, covered by leaf sheaths. **Leaves** 3–5, narrowly elliptic, 100–200 × 10–28mm. **Inflorescence** fairly dense, equalling or longer than the leaves, branched, many-flowered; inflorescence stalk covered in membranous sheaths. **Flowers** variable from yellow to yellow-green, pink or purple. Sepals unequal; median sepal narrowly triangular to oblong, 3–3.5 × 2mm; lateral sepals obliquely triangular, 4 × 3mm; petals strap- to spatula-shaped, 2.8–3 × 0.8mm; lip 3-lobed, 4–5mm long, disc with powdery hairs but without a callus; mentum conical, 2.5–3mm tall.

Distinguishable from *P. concreta* by the absence of an obvious callus on the disc of the lip.

| A | S | O | N | D | J | F | M | A | M | J | J |

Rare in riverine forest and bushveld, from near sea level to 500m. Flowers between December and February.

Polystachya ottoniana

Heidelberg (Western Cape) 12.11.2010

Hogsback 12.09.2013

Graskop 29.10.2013

Very slender epiphyte or lithophyte 40–150mm tall; pseudobulbs adjacent, often forming long chains, conical, 10–25 × 5–15mm. **Leaves** 2 or 3, linear to strap-shaped, 20–130 × 4–8mm. **Inflorescence** lax, as tall as the leaves, simple, hairy, 1–5-flowered. **Flowers** cream or white, lip with central yellow marking. Sepals subequal; median sepal narrowly oblong, 7.5–13 × 3–5mm; lateral sepals 9–13 × 4–7mm; petals narrowly spatula- to spear-shaped, 7–11 × 2–4mm; lip 3-lobed, 9–12 × 6–8.5mm, disc with a central ridge-like callus, surface papillate; mentum widely conical, 4–6mm tall.

A S O N D J F M A M J J

Common in a variety of forested habitats and often forming large mats on tree trunks and rocks, from sea level to 1,800m. Hybridizes with *P. transvaalensis*. Flowers between August and December.

Polystachya pubescens

Syn: *Polystachya ngomensis*

Umtamvuna 22.10.2011

Umtamvuna 22.10.2011

Umtamvuna 06.10.2012

Umtamvuna 22.10.2011

Slender epiphyte or lithophyte 100–200mm tall; pseudobulbs ovoid, closely aggregated, often forming long chains, 20–35 × 5–15mm. **Leaves** 2 or 3, strap-shaped to elliptic, green to deep purple, 25–120 × 9–20mm. **Inflorescence** lax, as tall as or taller than leaves, simple, 8–20-flowered; inflorescence stalk hairy. **Flowers** opening widely, deep yellow, lip and lateral sepals often with reddish markings. Sepals similar; median sepal narrowly oblong, 8–10 × 3–5mm; lateral sepals oblong, 8–11 × 3–6mm; petals egg-shaped, 6–9 × 2.5–6mm; lip 3-lobed, without a callus, 7 × 5–6mm, densely hairy, particularly on side lobes.

A	S	O	N	D	J	F	M	A	M	J	J

Fairly common in warm coastal forests, either growing epiphytically in forests, or on rocks in full sun, from near sea level to 1,500m. Flowers between August and December.

Polystachya sandersonii

Eshowe 09.11.2012

Ndulinde 25.10.2013

Ndulinde 27.10.2013

Ndulinde 25.10.2013

Eshowe 24.10.2012

Slender epiphyte 80–200mm tall; pseudobulbs adjacent, narrowly conical, 20–40 × 3–10mm. **Leaves** 2–4, strap-shaped, 40–100 × 7–20mm. **Inflorescence** dense, taller than the leaves, usually simple, 5–15-flowered; inflorescence and flower stalks hairy. **Flowers** with sepals brown, petals brownish-yellow and lip white, sometimes with maroon spots near the base. Sepals unequal, outer surface smooth or velvety; median sepal triangular, 7–9 × 3.2–5mm; lateral sepals obliquely egg-shaped, 9–12 × 5–7.5mm; petals spatula-shaped, 6.5–7 × 1.5–2.8mm; lip 3-lobed, erect at the base, disc with a central ridge; mentum conical, 4–5mm tall.

Distinguishable from *P. zambesiaca* by the smooth or velvety outer surface of the sepals and the maroon spots near the base of the lip. This species and *P. zambesiaca* are closely related and may perhaps be just one species.

A S O N D J F M A M J J

Fairly common in warmer forests, from near sea level to 1,400m. Flowers mostly between October and December.

Polystachya transvaalensis

Graskop 23.01.2012

Graskop 05.07.2013

Graskop 05.07.2013

Graskop 23.01.2012

Slender epiphyte 100–300mm tall; stems covered by leaf sheaths, not swollen into pseudobulbs. **Leaves** 3–6, leathery, strap-shaped to elliptic, rounded, 40–100 × 10–25mm. **Inflorescence** fairly dense, simple or rarely branched, shorter to slightly longer than the leaves, 5–10-flowered. **Flowers** with sepals cream, yellow, green or orange, petals and lip paler. Sepals unequal; median sepal narrowly triangular, 6–11 × 3.5–4.2mm; lateral sepals obliquely triangular, 8–13 × 7–10mm; petals spatula-shaped, 5–10 × 1.3–2mm; lip obscurely 3-lobed, upper surface hairy, 12 × 7mm, disc with a knob-like callus; mentum conical, 6.5–8.5mm tall.

A S O N D J F M A M J J

Subsp. *transvaalensis* is widespread in Africa and the only subspecies in the region.

Distinguishable from *P. albescens* by the leathery, rounded leaves and inflorescence with fewer than 10 flowers.

Locally common in cool, moist montane forests, from 900–1,900m. Hybridizes with *P. ottoniana*. Flowers between August and February.

Polystachya zambesiaca

Tzaneen 12.05.2013

Tzaneen 12.09.2013

Tzaneen 06.10.2013

Slender epiphyte 80–170mm tall; pseudobulbs narrowly conical, 25–30 × 7–10mm. **Leaves** 2 or 3, strap- to spear-shaped, 70–130 × 7–14mm. **Inflorescence** dense, as tall as or occasionally taller than the leaves, simple, 5–16-flowered; inflorescence and flower stalks hairy. **Flowers** yellow, lip white with violet streaks on inner surface of side lobes. Sepals densely hairy on the outside; median sepal narrowly triangular, 6.5–12 × 3.5–5mm; lateral sepals obliquely narrowly triangular, 8–13 × 3–6mm; petals spear-shaped, 5–10 × 1.5–3mm; lip 3-lobed, 5–8 × 4–6mm, disc with a powdery white, slightly raised callus; mentum 5–6mm tall.

Distinguishable from *P. sandersonii* by the hairy outer surface of the sepals and the lip side lobes streaked with violet. This species and *P. sandersonii* are closely related and may perhaps be just one species.

| A | S | O | N | D | J | F | M | A | M | J | J |

Fairly common in the warmer forests, from 1,000–1,400m. Flowers between August and February.

Polystachya zuluensis

Mbabane (Swaziland) 09.05.2013

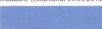

Mbabane (Swaziland) 09.05.2013

Mbabane (Swaziland) 09.05.2013

Mbabane (Swaziland) 09.05.2013

Slender epiphyte or lithophyte up to 400mm tall; pseudobulbs ovoid, clustered, 20–30 × 6–15mm. **Leaves** 2 or 3, deciduous, often dried at flowering time, curved, linear, 55–150 × 5–7mm. **Inflorescence** wiry, branched, most flowers borne in terminal segment. **Flowers** rarely fully open, pale lilac, lip with a central yellow streak. Sepals similar; median sepal strap-shaped to triangular, 8–11 × 2.2–3mm; lateral sepals narrowly spear-shaped, 7–11 × 3–4mm; petals narrowly spear-shaped, 7–10 × 2.5–3mm; lip egg-shaped, obscurely 3-lobed, 6.5–9 × 4.5–6mm, disc either without a callus or with a small central ridge; mentum obscure, 1.8–3mm tall.

A S O N D J F M A M J J

Restricted but often abundant as a lithophyte or an epiphyte on *Xerophyta retinervis*, from 1,200–1,300m. Flowers between March and July.

Pterygodium acutifolium

Grabouw 27.11.2009

Betty's Bay 30.10.2008

Rooiels 30.10.2006

Cape Peninsula 28.11.2007

Slender terrestrial up to 470mm tall. **Leaves** 3 or 4, spaced up the stem, oblong, up to 180 × 23mm. **Inflorescence** lax to fairly dense, 1–14-flowered. **Flowers** with sepals lime-green, petals and lip rich yellow; 17 × 15mm. Sepals 8–12 × 4–6mm; median sepal fused to the petals; lateral sepals spreading; petals 12–13 × 8–10mm; lip 5 × 3–4mm, margin wavy; appendage triangular, apex smooth and down-curved.

Distinguishable from *P. catholicum* by the generally richer butter-yellow flower colour, the smooth appendage apex and older flowers that do not turn red.

A S O N D J F M A M J J

Common and widespread in swamps, on rock sheets and in seepages, from near sea level to 1,500m. Pollinated by oil-collecting bees. Flowers between late October and December, after fire.

Pterygodium alatum

Op-die-Berg 20.09.2009

Bredasdorp 28.08.2010

Ceres 05.10.2010

Piketberg 16.09.2008

Vanrhynsdorp 11.09.2006

Slender terrestrial up to 150mm tall. **Leaves** numerous, narrowly elliptic, up to 60 × 10mm. **Inflorescence** fairly dense, 2–many-flowered. **Flowers** pale greenish-yellow; 24 × 15mm. Median sepal fused to the petals, 7–11 × 2–3mm; lateral sepals spreading to reflexed, 9–10 × 3mm; petals 7–10 × 6–7mm; lip with 2 broad, diverging lobes, each 5–7 × 2–5mm, and a small pointed midlobe 1.5mm long; appendage spatula-shaped, with a hood-like gland apically.

A	S	O	N	D	J	F	M	A	M	J	J

Common along the coastal forelands and foot of the mountain ranges, from sea level to 1,000m. Pollinated by oil-collecting bees. Flowers between August and October, stimulated by fire.

Pterygodium caffrum

Paarl 26.09.2009

Stanford 17.11.2009

Grabouw 29.10.2006

Houwhoek 10.11.2010

Slender to fairly robust terrestrial up to 250mm tall in the early flowering form and 380mm tall in the late flowering form. **Leaves** 4 or more, cauline, egg- to spear-shaped, up to 85 × 32mm. **Inflorescence** fairly dense, few- to many-flowered; bracts reaching to halfway up the flowers. **Flowers** with sepals pale greenish, petals and lip yellow; 28 × 15mm. Median sepal fused to the petals, 9–11 × 2.5–3mm; lateral sepals spreading, 9–11 × 5 × 2mm; petals 7–12 × 5–7mm; lip with 2 broad, diverging, rounded lobes, each 7–9 × 5–8mm; appendage erect and rather stout, apex with large ventral cavity.

Distinguishable from *P. pentherianum* by the larger, yellow flowers.

A S O N D J F M A M J J

Locally common in fynbos, from near sea level to 500m. Pollinated by oil-collecting bees. Flowers between August and December, with distinct peaks in September (early flowering) and November (late flowering), stimulated by fire.

Pterygodium catholicum

Betty's Bay 17.11.2008

Grabouw 25.10.2006

Hermanus 22.10.2006

Ceres 08.10.2009

Slender terrestrial up to 300mm tall in the early flowering form and 450mm tall in the late flowering form. **Leaves** 2–4, oblong to spear-shaped, up to 120 × 30mm. **Inflorescence** lax, 1–14-flowered. **Flowers** with sepals green, petals and lip yellowish-green, sometimes flushed with red and older flowers turning completely red; 25 × 15mm. Median sepal fused to the petals, 9–10 × 3.5–4mm; lateral sepals spreading, 7–13 × 4–5mm; petals 9–12 × 9mm; lip 4–6 × 2.5–3mm; appendage erect, finely toothed and curved forward apically.

Distinguishable from *P. acutifolium* by the generally yellowish-green flower colour, the finely toothed appendage apex and older flowers that turn red.

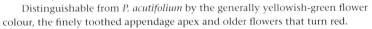

A S O N D J F M A M J J

Common and ubiquitous in open short scrub, from near sea level to 1,600m. Pollinated by oil-collecting bees. Flowers between August and December, with distinct peaks in September (early flowering) and November (late flowering), after fire in fynbos, but not fire-dependent in renosterveld.

Pterygodium cleistogamum

Syn: *Pterygodium newdigateae* var. *cleistogamum*

Tsitsikamma 03.12.2007

Tsitsikamma 05.11.2012

Slender terrestrial up to 330mm tall. **Leaves** 2–4, along the stem, elliptic to oblong, up to 70 × 20mm. **Inflorescence** lax, 3–7-flowered. **Flowers** greenish-yellow, bud-like and not opening, 30 × 10mm. Median sepal fused to the petals, 12 × 5mm; lateral sepals spreading, 8–11 × 4mm; petals 12mm long; lip 12mm long, lacking appendage.

A S O N D J F M A M J J

Found on stony slopes, from near sea level to 340m. Sets seed without pollination. Flowers between November and December, stimulated by fire.

Pterygodium cooperi

Naudesnek Pass 03.02.2014

Naudesnek Pass 07.02.2008

Naudesnek Pass 17.02.2008

Naudesnek Pass 08.02.2008

Slender terrestrial up to 290mm tall. **Leaves** 2–4, the basal leaf generally considerably larger, broadly egg-shaped to oblong, up to 160 × 50mm; **cauline leaves** spear-shaped to elliptic. **Inflorescence** lax, 5–many-flowered. **Flowers** with sepals pale green to purplish-green, petals white and green, lip and appendage white; 20 × 15mm. Median sepal fused to the petals, 6–8 × 2–4mm; lateral sepals reflexed, 7–9 × 4–5mm; petals 6–9 × 6–11mm; lip 3–5mm long, with 2 stout, deflexed side lobes; appendage triangular, ridged, 4–5 × 5mm.

A S O N D J F M A M J J

Localized at the upper margin of montane forest or at higher altitudes below big rocks and cliffs or on steep, wet grassland slopes, from 1,500–2,800m. Pollinated by oil-collecting bees. Flowers between January and March.

Pterygodium cruciferum

Durbanville 14.10.2010

Durbanville 14.10.2010

Durbanville 11.10.2009

Darling 06.10.2007

Slender to fairly robust terrestrial up to 400mm tall. **Leaves** 2 or 3, cauline, linear to elliptic, up to 190 × 18mm. **Inflorescence** lax, 2–10-flowered. **Flowers** yellowish-green, 35 × 30mm. Median sepal fused to the petals, 18–22 × 4–6mm; lateral sepals spreading, 12–15 × 8mm; petals 15–18 × 11–14, 3mm deep; lip very narrowly linear, 3–4mm long; appendage cross-shaped, 8–14mm tall, with horizontal lobes broad, 5mm long.

| A | S | O | N | D | J | F | M | A | M | J | J |

Rare in fynbos and renosterveld, from near sea level to 240m. Flowers between September and November, stimulated by fire.

Pterygodium hallii

Nieuwoudtville 22.09.2009

Nieuwoudtville 22.09.2009

Nieuwoudtville 05.09.2006

Nieuwoudtville 08.10.2007

Slender to robust terrestrial up to 500mm tall. **Leaves** 6–many, sheathing, spear-shaped to narrowly elliptic, up to 180 × 35mm. **Inflorescence** lax to fairly dense, many-flowered. **Flowers** non-resupinate, mostly green, lip densely spotted dark purple; 30 × 15mm. Median sepal fused to the petals, 7–10 × 3–4mm; lateral sepals reflexed, 10–13 × 2–3mm; petals rounded, 8–14 × 6–10mm; lip 4–6 × 4–8mm; appendage oblong-rounded, 6–8 × 4–5mm, with 2 keels basally.

Distinguishable from *P. inversum* by the appendage apex that is never 2-lobed and has a maroon to purple-black patch at the base.

| A | S | O | N | D | J | F | M | A | M | J | J |

Occasional in short dry scrub, from sea level to 1,600m. Pollinated by oil-collecting bees. Flowers between September and October.

Pterygodium hastatum

Sabie 18.02.2011

Witsieshoek 08.02.2007

Garden Castle 02.02.2009

Sabie 02.02.2011

Giant's Castle 04.02.2007

Slender terrestrial up to 350mm tall. **Leaves** 2, cauline, egg-shaped to oblong, up to 210 × 60mm. **Inflorescence** lax, 2–8-flowered. **Flowers** pale green with small purple dots on petal and median sepal apices, 20 × 15mm. Median sepal fused to the petals, 8–14 × 4–5mm; lateral sepals spreading, 7–12 × 3–4mm; petals almost circular, margin somewhat wavy, 6–10 × 8–12mm; lip 3 × 3–5mm; appendage 3-pointed apically, 4–7 × 3mm.

A S O N D J F M A M J J

Sporadic in montane to subalpine grassland, from 1,200–3,000m. Pollinated by oil-collecting bees. Flowers between January and March.

Pterygodium inversum

Syn: *Anochilus inversus*

Mamre 25.09.2009

Botrivier 27.09.2009

Mamre 03.10.2008

Fairly robust terrestrial up to 550mm tall. **Leaves** 6–many, cauline, spear-shaped, up to 200 × 40mm. **Inflorescence** dense, many-flowered. **Flowers** non-resupinate, pale green, 25 × 12mm. Median sepal fused to the petals, 8–9 × 2.5mm; lateral sepals 8–10 × 3mm; petals rounded, 8–12 × 5–8mm; lip rounded, 3–4 × 3–4mm; appendage egg-shaped to oblong, 4–5 × 3–4mm, distinctly 2-lobed apically and with 2 projections basally.

Distinguishable from *P. hallii* by the appendage apex that is 2-lobed and without a maroon to purple-black patch at the base.

A	S	O	N	D	J	F	M	A	M	J	J

Rare in renosterveld, mostly on the coastal forelands, from 90–1,000m. Pollinated by oil-collecting bees. Flowers between September and October, stimulated by fire.

Pterygodium leucanthum

Naudesnek Pass 04.03.2012

Naudesnek Pass 03.02.2014

Sani Pass 25.01.2010

Naudesnek Pass 04.03.2012

Slender terrestrial up to 350mm tall. **Leaves** 2–10, cauline, egg- to spear-shaped, up to 170 × 35mm. **Inflorescence** fairly lax, 3–15-flowered. **Flowers** with sepals green, petals cream, lip appendage pale green with dark green apex; 18 × 12mm. Median sepal fused to the petals, 8–9 × 3mm; lateral sepals spreading, pouch-shaped, 7–8 × 2mm deep; petals almost circular, 6–7 × 8–9mm; lip reduced to 2 semi-circular lobes, 1–1.5mm long, with a minute central projection; appendage 6 × 3mm, topped by 2 rounded, down-curved, fleshy lobes.

A S O N D J F M A M J J

Occasional in montane grassland, from 1,300–2,400m. Pollinated by oil-collecting bees. Flowers between January and early March.

Pterygodium magnum

Syn: *Corycium magnum*

Dullstroom 31.01.2011

Ngeli 16.01.2014

Ngeli 27.02.2010

Ngeli 29.01.2009

Very robust terrestrial up to 1.5m tall. **Leaves** 6–many, oblong to spear-shaped, up to 250 × 60mm. **Inflorescence** dense, 45–100-flowered; bracts much longer than flowers, deflexed, green at flowering. **Flowers** with sepals green, petals yellowish, with red dots and darker veins, lip pale green to mauve with darker veins; 25 × 12mm. Median sepal fused to the petals, 8–12 × 2.5–5mm; lateral sepals spreading, 7–12 × 3–5mm; petals 8–12 × 5–7mm, margins fringed; lip 6–8 × 4–6mm, with base 1mm wide, margin fringed; appendage 2-lobed and pouch-shaped in upper half, 7–9 × 3mm.

A	S	O	N	D	J	F	M	A	M	J	J

Occasional along forest edges and in grassland, from 1,200–2,000m. Pollinated by oil-collecting bees. Flowers between January and March.

Pterygodium pentherianum

Ceres 19.09.2007

Citrusdal 28.08.2007

Citrusdal 09.09.2008

Citrusdal 28.08.2007

Slender terrestrial up to 200mm tall. **Leaves** 1–3, cauline, oblong to spear-shaped, up to 80 × 25mm. **Inflorescence** lax, 4–8-flowered. **Flowers** white to pale green, 25 × 15mm. Median sepal fused to the petals, 8–10 × 3mm; lateral sepals spreading, 9–11 × 4–6mm; petals 10–11 × 6mm, outer margin notched with small, regular teeth; lip 5–7 × 14mm, with 2 diverging, corky, almost circular lobes 4.5mm in diameter; appendage erect, fleshy, broad, 3-lobed apically, 4 × 4mm.

Distinguishable from *P. caffrum* by the smaller, white or pale green flowers.

A S O N D J F M A M J J

Found in clay soil among short scrub in renosterveld, from 150–1,250m. Pollinated by oil-collecting bees. Flowers between August and September.

Pterygodium platypetalum

Ceres 19.09.2008

Ceres 19.09.2008

Ceres 02.09.2010

Ceres 19.09.2007

Slender terrestrial up to 150mm tall. **Leaves** usually 2, narrowly elliptic, up to 60 × 10mm. **Inflorescence** 1- or 2-flowered. **Flowers** pale yellowish-green, 18 × 15mm. Median sepal fused to the petals, 9 × 3.5mm; lateral sepals spreading, apex recurved, 8–11 × 4–5mm; petals 8–12 × 10–15mm; lip with 2 small rounded side lobes, 3–4mm long; appendage triangular, central lobe reflexed, lateral lobes erect, 4–5mm long.

A S O N D J F M A M J J

Occasional on mountain slopes, often on ledges or in shallow soils, from 80–1,500m. Flowers between September and October, stimulated by fire.

Pterygodium schelpei

Matjiesfontein 18.09.2013

Clanwilliam 06.09.2010

Clanwilliam 31.08.2010

Clanwilliam 06.09.2010

Slender terrestrial up to 300mm tall. **Leaves** 2 or 3, basal leaf the largest, strap-shaped to narrowly egg-shaped, 80–150 × 25–40mm, upper leaves narrower, tapering towards apex. **Inflorescence** lax, 8–15-flowered; bracts deflexed. **Flowers** with sepals green, petals and lip greenish-white, often fully white; 5.8–10 × 9–12mm. Median sepal fused to the petals, 6–7 × 1.5–2.5mm; lateral sepals spreading to reflexed, 5.5–6.5 × 4–4.5mm; petals 2-lobed, axe-shaped; lip 3-lobed, hanging; lip midlobe and side lobes 3–4mm long, basally thickened, raised area shorter than 1mm; appendage erect, cup-shaped.

   Distinguishable from *P. volucris* by the white flowers, the significant lip side lobes and the minute outgrowths of the lip.

Localized in seepages or cool habitats on clay soils, often under bushes or in the lee of rocks, from 500–1,050m. Pollinated by oil-collecting bees. Flowers between late August and October.

452   *Pterygodium*

Pterygodium vermiferum

Gansbaai 21.09.2011

Gansbaai 21.09.2011

Gansbaai 05.10.2011

Gansbaai 14.09.2006

Slender terrestrial up to 170mm tall. **Leaves** 1–3, cauline, basal leaf the largest, prostrate, spear- to egg-shaped, up to 90 × 40mm, often dry at flowering, upper leaves narrower. **Inflorescence** lax to fairly dense, 12–17-flowered; bracts deflexed. **Flowers** green to yellowish-green becoming yellow, lip whitish; 10 × 9mm. Median sepal fused to the petals, 5 × 3mm; lateral sepals 4 × 3.5–4mm; petals 5 × 2.5mm; lip with lateral ring-like projections and 2 lateral thickened, raised areas; appendage erect; rostellum with worm-like outgrowths.

A S O N D J F M A M J J

Very rare in strandveld and limestone fynbos, from 50–385m. Pollinated by oil-collecting bees. Flowers between September and early October, stimulated by fire.

Pterygodium volucris

Piketberg 30.08.2010

Piketberg 30.08.2010

Paarl 06.09.2010

Paarl 06.09.2010

Slender terrestrial up to 350mm tall. **Leaves** 3 or 4, cauline, egg-shaped to oblong, up to 140 × 50mm. **Inflorescence** lax to dense, 10–30-flowered; bracts deflexed. **Flowers** with sepals green, petals and lip lime-green; 15 × 12mm. Median sepal fused to the petals, 7–8 × 2mm; lateral sepals reflexed, 6–7 × 2.5–3mm; petals folded apically, 7–7.5 × 5–6mm; lip 4–5 × 6mm, with rounded side lobes spreading 1mm laterally, basally thickened, raised area mostly reaching beyond the lip; appendage spatula- to shield-shaped, forming a cup, apically notched, 3–4 × 2mm.

Distinguishable from *P. schelpei* by the lime-green flowers, the insignificant lip side lobes, and the outgrowths of the lip normally overlapping the lip side lobes.

A S O N D J F M A M J J

Common in sandy or clay soils, often seen under bushes, from near sea level to 900m. Pollinated by oil-collecting bees. Flowers between late August and October, stimulated by fire.

Rangaeris muscicola

Umtamvuna 19.12.2011

Umtamvuna 13.01.2014

Umtamvuna 19.12.2011

Umtamvuna 13.01.2014

Slender to robust lithophyte or epiphyte; stems short and stout, 30–80mm long, 7–10mm in diameter. **Leaves** folded together, strap-shaped, fleshy or leathery, arranged in a fan, 60–110 × 12–20mm. **Inflorescences** usually several, lax, usually shorter than the leaves, 55–150mm long, 5–15-flowered. **Flowers** white, turning orange with age. Sepals and petals similar, 7–11 × 3.5–6mm; lip rhomboid, 10–12 × 8mm; spur 50–90mm long.

A S O N **D J** F M A M J J

Occasional on trees in riverine forest or exposed on sandstone, from 100–1,100m. Flowers between December and January.

Rhipidoglossum xanthopollinium

Syn: *Diaphananthe xanthopollinia*

Eshowe 15.02.2013

Eshowe 15.02.2013

Hluleka Nature Reserve 31.01.2013

Hluleka Nature Reserve 31.01.2013

Slender to robust epiphyte; stems frequently branched, up to 150mm long; roots produced opposite the leaves, up to 300mm long, branching. **Leaves** leathery, strap-shaped, unequally bilobed, 35–80 × 4–10mm. **Inflorescences** several per stem, lax to semi-dense, 30–70mm long, 6–25-flowered. **Flowers** greenish-yellow, translucent. Sepals unequal; median sepal oblong, 3–3.5 × 1.5–2mm; lateral sepals strap-shaped, 3–4 × 1–1.5mm; petals broadly elliptic, 2.5–3 × 2.5mm; lip 3–4 × 5–6mm; spur curved forward, 4.5–6mm long.

| A | S | O | N | D | J | F | M | A | M | J | J |

Locally common in subtropical low-altitude forests, from near sea level to 1,000m. Flowers between late January and March.

Satyrium acuminatum

Joubertina 01.10.2010

Tsitsikamma 05.11.2007

Tsitsikamma 04.11.2012

Nature's Valley 31.08.2013

Tsitsikamma 04.11.2012

Robust to slender terrestrial 150–350mm tall. **Leaves** 2, pressed flat on the ground, egg-shaped to almost circular, 20–80mm long. **Inflorescence** moderately dense, 15–27-flowered; bracts partly to fully deflexed. **Flowers** white to pale pink. Sepals and petals similar, free to the base, basal two thirds to three quarters forming a broadly U-shaped entrance to the galea, apices deflexed, 4mm long; lip viewed from the front as deep as high, entrance 4–7mm high, flap 1mm long and partly reflexed; spurs slender, 12–22mm long, close to ovary.

| A | S | O | N | D | J | F | M | A | M | J | J |

Occasional to locally frequent in moist fynbos or grass and bush vegetation on open south-facing slopes, mainly from 200–500m and occasionally as low as sea level and as high as 1,600m. Hybridizes with *S. ligulatum*. Flowers between September and December, stimulated by fire.

Satyrium bicallosum

Ceres 09.10.2009

Villiersdorp 07.10.2008

Ceres 11.10.2009

Slender to stout terrestrial 120–220mm tall. **Leaves** 1–3, near soil surface, up to 12–65mm long. **Inflorescence** 6–150-flowered; bracts spreading then bending upward, 2–10 times the length of the ovary. **Flowers** dull white, faintly greenish, with a pale purple patch above the entrance to each spur. Median sepal 1–3mm long; lateral sepals 2–4mm long; petals slightly shorter than median sepal; lip hooded, forming a widely 2-lobed hood, flap bent down to almost touching median sepal and dividing floral entrance into two; spurs 1mm long.

Distinguishable from *S. retusum* by the lip apex that is folded downward in front of column.

A S O N D J F M A M J J

Locally common on dry to damp sandy soils, from near sea level to 1,000m. Pollinated by fungus gnats. Hybridizes very rarely with *S. candidum*. Flowers between October and November, after fire.

Satyrium bicorne

Cape Peninsula 21.09.2008

Paarl 04.09.2010

Vanrhynsdorp 11.09.2006

Piketberg 16.09.2008

Slender terrestrial 190–380mm tall. **Leaves** usually 2, pressed flat on the ground, egg-shaped to almost circular, 30–180mm long, with cup-like leaf sheaths higher up. **Inflorescence** dense to lax, 4–40-flowered; bracts deflexed, 1.4–2.4 times the length of the 6–14mm-long ovary. **Flowers** pale greenish-yellow, faintly to darkly tinged with purple-brown. Sepals and petals united with the lip to form a tube for a third to half their length, then abruptly deflexed; sepals 6–9mm long; petals slightly shorter; lip with entrance facing partly downward, flap not bent back; spurs 10–22mm long.

A S O N D J F M A M J J

Common in sandy soils, growing in open to partly sheltered places among bushes, mainly from 200–500m and occasionally as low as sea level or as high as 1,200m. Pollinated by moths. Hybridizes with *S. erectum*. Flowers between September and October, stimulated by fire.

Satyrium bracteatum

Cape Peninsula 20.09.2008

Cape Peninsula 22.10.2009

Naudesnek Pass 11.12.2008

Verloren Vallei 17.11.2013

Storms River Mouth 06.10.2010

Slender to stout terrestrial 80–160mm tall. **Leaves** 3–8, up to 60mm long, gradually getting smaller up the stem. **Inflorescence** fairly dense, 4–15-flowered; bracts generally overtopping flowers. **Flowers** yellow with dark reddish-brown stripes. Median sepal 4–6mm long, fused to bases of petals and lateral sepals, the sepals slightly longer with the free part bent down; lip taller than deep; spurs pouch-shaped, shorter than 1mm. Very variable across its range.

A S O N D J F M A M J J

Locally common in a variety of habitats ranging from dry flats to wet marshy sites, peaty ledges and fissures amongst rocks, from near sea level to 2,500m. Pollinated by flies. Flowers between September and December, stimulated by fire.

Satyrium candidum

Kleinmond 06.10.2011

Kleinmond 03.10.2011

Franschhoek 08.11.2007

Franschhoek 02.11.2008

Stout terrestrial 230–450mm tall. **Leaves** 2, pressed flat on the ground, broadly egg-shaped to almost circular, up to 130mm long, with cup-like leaf sheaths higher up. **Inflorescence** rather lax, 14–23-flowered; bracts partly to fully deflexed, 1.2–2.5 times the length of the 8–18mm-long ovary. **Flowers** white to ivory, faintly tinged with pink. Sepals fused for one third of their length with the petals, then bent down, 9–12mm long; lip with entrance 4–5mm high, flap bent backward, 2mm long; spurs 10–20mm long.

A S O N D J F M A M J J

Rare on sandy flats and on mountains, from near sea level to 1,500m. Hybridizes very rarely with *S. bicallosum.* Flowers between September and November, stimulated by fire.

Satyrium carneum

Gansbaai 12.10.2010

Gansbaai 12.10.2010

Gansbaai 12.09.2006

Gansbaai 17.10.2007

Robust terrestrial 370–710mm tall. **Leaves** 2–4, thick and fleshy, lowest 2 partly pressed onto the ground, 70–230mm long, with gradual transition to the sheaths higher up. **Inflorescence** dense, stout, 19–38-flowered; bracts partly deflexed, 1.4–2.3 times the length of the 11–18mm-long ovary. **Flowers** pale pink to rose, very rarely white. Sepals and petals almost free, rolled back apically, 13–18mm long; lip entrance 12–14mm high, flap 2mm long, completely bent backward; spurs 14–20mm long.

A S O N D J F M A M J J

Increasingly local and rare among dune-bush vegetation, in fynbos on coastal hills and ridges, on moist to dry sands and limestones, from near sea level to 600m. Pollinated by sunbirds. Hybridizes with *S. coriifolium*. Flowers between September and November, stimulated by fire.

Satyrium coriifolium

Ceres 20.09.2007

Greyton 12.10.2007

Cape Peninsula 20.08.2010

Paarl 17.10.2009

Fairly robust terrestrial 230–480mm tall. **Leaves** 2–4, erect to spreading, purple-spotted towards the sheathing leaf bases, rather thick and stiff, elliptic to egg-shaped, 30–150mm long. **Inflorescence** 7–20-flowered; bracts deflexed, 1.7–3.3 times the length of the 10–14mm-long ovary. **Flowers** bright yellow to bright orange, lip often tinged with red. Sepals and petals fused in basal quarter to third, 7–13mm long; lip as deep as tall, flap 2mm long, bent backward; spurs 9–12mm long.

| A | S | O | N | D | J | F | M | A | M | J | J |

Locally frequent in sandy, moist, open flats, from near sea level to 750m. Pollinated by sunbirds. Hybridizes with *S. erectum*, *S. acuminatum*, *S. pallens* and *S. carneum*. Flowers between August and October.

Satyrium cristatum

Slender terrestrial 240–550mm tall. **Leaves** 2, rarely 3, spreading near the ground, egg-shaped to elliptic, up to 160mm long. **Inflorescence** moderately dense, 15–60-flowered; bracts fully deflexed, 2–3 times the length of the 4–9mm-long ovary. **Flowers** white, sepals, petals and inside of lip with bright purple-red to brownish-red blotches and lines. Sepals 6–10mm long, variously fused to petals and lip; lip entrance 3–7mm high, flap 0.7mm long, bent backward; spurs 3–12mm long.

Satyrium cristatum var. *cristatum*

Graskop 10.03.2012

Sabie 27.01.2014

Sabie 27.01.2014

Graskop 29.01.2014

Naudesnek Pass 24.02.2010

Var. *cristatum* is distinguished as follows: lip entrance 3–6mm tall; spurs 3–7mm long, usually shorter than ovary; rostellum notched, with spherical viscidia.

Var. *cristatum* occurs in the eastern part of South Africa, as well as in Lesotho and Swaziland.

Common on moist grassy flats, sometimes in marshy conditions, from near sea level to 2,200m. Flowers between January and March.

A S O N D J F M A M J J

Satyrium cristatum var. *longilabiatum*

Dullstroom 27.01.2012

Sabie 07.02.2012

Sabie 27.01.2014

Sabie 27.01.2014

Var. *longilabiatum* is distinguished as follows: lip entrance 5–7mm tall; spurs 7–12mm long, usually longer than ovary; rostellum rounded, with plate-like viscidia resembling rabbit teeth.

Var. *longilabiatum* occurs in the eastern part of South Africa, as well as in Swaziland.

Less common but often growing together with var. *cristatum* in marshy grassland, mostly from 1,000–2,000m. Flowers between January and March, with a peak in January.

| A | S | O | N | D | J | F | M | A | M | J | J |

Satyrium emarcidum

Bredasdorp 28.09.2010

Gansbaai 13.09.2006

Gansbaai 31.10.2009

Gansbaai 04.10.2011

Gansbaai 27.10.2009

Stout to slender terrestrial 300–450mm tall. **Leaves** 2 or 3, basal leaf pressed flat on the ground, broadly egg-shaped to almost circular, up to 80mm long. **Inflorescence** moderately dense, 10–20-flowered; bracts completely deflexed. **Flowers** greenish-yellow, tips of petals and sepals often flushed with maroon-red and soon drying brown after the flowers open. Sepals 10mm long, ascending; petals similar to lateral sepals but shorter, ascending; lip hooded, flap prominent, reflexed; spurs 9–10mm long.

Distinguishable from *S. ligulatum* by the dirty greenish-yellow flower colour and the broadly egg-shaped basal leaf pressed flat on the ground.

A S O N D J F M A M J J

Localized but often forming large populations in dunes and on coastal sand flats, from sea level to 170m. Flowers between September and October, stimulated by fire.

Satyrium erectum

Citrusdal 30.08.2011

Nieuwoudtville 06.09.2006

Nieuwoudtville 15.08.2006

Nieuwoudtville 14.08.2006

Slender to robust terrestrial 200–420mm tall. **Leaves** 2, pressed flat on the ground, egg-shaped to elliptic, 40–160mm long. **Inflorescence** dense, 11–37-flowered; bracts partly to fully deflexed, 1.8–3.5 times the length of the 6–14mm-long ovary. **Flowers** usually pale to deep pink, petals with darker tinges and spots. Sepals and petals fused to sides of lip for a third of their length, then gradually curving downward; lip entrance 6–8mm high, flap 3mm long, only slightly bent backward; spurs 5–11.5mm long, usually shorter than ovary.

Distinguishable from *S. pallens* and *S. pulchrum* by the shorter spurs, the flowers with spots and stripes and the pungent odour.

Common in areas with generally rather low rainfall (200–400mm per annum), in karroid scrub, renosterveld and drier forms of fynbos, usually from 100–700m, occasionally up to 1,500m. Pollinated by bees. Hybridizes with *S. coriifolium*, *S. bicorne*, *S. pallens* and *S. humile*. Flowers between August and October, stimulated by fire.

Satyrium eurycalcaratum

Oudtshoorn 17.12.2009

Oudtshoorn 20.12.2013

Oudtshoorn 20.12.2013

Oudtshoorn 17.12.2009

Moderately robust terrestrial 155–355mm tall. **Leaves** 2, flat on the ground, withered at flowering time, broadly egg-shaped, up to 42mm long. **Inflorescence** lax, 2–17-flowered; bracts deflexed. **Flowers** white, tinged with purplish-pink to red-brown. Petals and sepals projecting forward for half their length, then gradually reflexing downward; lip entrance 4–4.8mm high, flap 3mm long, prominent, erect; spurs diverging, 21–26mm long.

Distinguishable from other long-spurred, white-flowered *Satyrium* species by the diverging spurs and the leaves that are withered at the time of flowering.

A S O N D J F M A M J J

Localized and rare in fynbos, on mountain slopes, from 1,200–1,800m. Flowers between December and January, stimulated by fire.

Satyrium foliosum

Hermanus 15.01.2014

Hermanus 15.01.2014

Hermanus 15.01.2014

Hermanus 15.01.2014

Slender to robust terrestrial 230–310mm tall. **Leaves** 2 or 3, more or less erect and clustered near the base, egg-shaped, 40–100mm long. **Inflorescence** dense, 4–30-flowered; bracts deflexed, 1.2–1.5 times the length of the 12–16mm-long ovary. **Flowers** pale yellowish-green on opening but soon changing to orange- or purple-brown. Sepals and petals fused with the lip in basal quarter, strongly recurving soon after flowering; median sepal and petals similar, lateral sepals slightly larger, 5–6mm long; lip entrance 3–4mm high, flap 1–2mm long, bent backward; spurs 20–25mm long.

| A | S | O | N | D | J | F | M | A | M | J | J |

Rare and confined to the summer mist-belt, in black peaty soil, from 600–1,600m. Flowers between January and February, after fire.

Satyrium hallackii

Robust terrestrial 200mm–1m tall. **Leaves** 4–6, cauline, partly erect, with a gradual transition to the sheaths, narrowly oblong to elliptic-oblong, 70–200mm long. **Inflorescence** dense to rather well spaced, 20–70-flowered; bracts slightly to partly to completely deflexed, 1–3 times the length of the 9–12mm-long ovary. **Flowers** pink, rarely white. Sepals fused to petals and lip in basal quarter, 6.5–10.5mm long; petals similar; lip entrance rounded, flap 1mm long, bent backward; spurs 7–31mm long.

Satyrium hallackii subsp. hallackii

Betty's Bay 04.01.2011

Pringle Bay 22.12.2013

Betty's Bay 05.12.2007

Subsp. *hallackii* is distinguished as follows: spurs 7–14mm long.
  Subsp. *hallackii* is restricted to the Western and Eastern Cape of South Africa.

Rare and local in moist and sometimes saline soils just inland from the shoreline, near sea level. Pollinated mainly by bees. Flowers between November and January.

A	S	O	N	D	J	F	M	A	M	J	J

Satyrium hallackii subsp. *ocellatum*

Syn: *Satyrium ocellatum*

Dullstroom 03.01.2011

Verloren Vallei 05.02.2013

Dullstroom 13.01.2010

Dullstroom 12.02.2010

Subsp. *ocellatum* is distinguished as follows: spurs 15–31mm long.

Subsp. *ocellatum* occurs from the Eastern Cape northwards to Malawi, including Lesotho and Swaziland.

Locally frequent in marshes, among tall grassy vegetation, from 600–2,200m. Pollinated mainly by hawkmoths. Flowers between December and March.

| A | S | O | N | D | J | F | M | A | M | J | J |

Satyrium humile

Houwhoek 15.10.2010

Ceres 12.10.2009

Franschhoek 13.09.2006

Citrusdal 31.10.2007

Franschhoek 13.09.2006

Slender terrestrial 150–280mm tall. **Leaves** usually 2, spreading near the ground, egg-shaped, 20–100mm long. **Inflorescence** dense, 10–20-flowered; bracts deflexed, 1.3–2 times the length of the 7–11mm-long ovary. **Flowers** pale cream tinged with green, pink or brownish. Sepals and petals fused to the lip in their basal quarter; sepals 6–9mm long; petals slightly smaller; lip viewed from the front deeper than tall, flap 2mm long, bent slightly backward; spurs 12–26mm long, standing away from ovary.

A S O N D J F M A M J J

Locally common in fynbos and renosterveld, on stony soil, from 150–1,500m. Hybridizes with *S. erectum*. Flowers between September and November but mainly in October, stimulated by fire.

Satyrium jacottetiae

Joubertina 20.09.2011

Joubertina 02.10.2010

Joubertina 20.10.2010

Joubertina 20.10.2010

Slender to robust terrestrial 230–480mm tall. **Leaves** 2, pressed flat on the ground, broadly egg-shaped to elliptic, 20–120mm long; sheaths above dry and membranous. **Inflorescence** fairly dense, 8–22-flowered; bracts dry and membranous, gradually becoming deflexed, 1.3–1.9 times the length of the 9–18mm long ovary. **Flowers** white to pale pink. Sepals and petals fused to the lip in the basal quarter; sepals 9–11mm long; petals fringed with thick hairs; lip 8–11mm high, 5–6mm deep, flap 3mm long, upright, fringed with thick hairs; spurs 21–27mm long.

Distinguishable from *S. membranaceum* by the short, broad rostellum, blunt in front from where it presents egg-shaped to triangular viscidia.

A	S	O	N	D	J	F	M	A	M	J	J

Occasional on grassy, open slopes, from 600–1,500m. Flowers between late September and October, stimulated by fire.

Satyrium ligulatum

Ruiterbos 03.11.2007

Ruiterbos 07.11.2007

Ceres 02.12.2008

Cape Peninsula 30.10.2006

Stout to slender terrestrial 170–550mm tall. **Leaves** 3–6, more or less erect and clustered near the base, egg-shaped, 30–140mm long. **Inflorescence** slender, 10–35-flowered; bracts deflexed, 1.5–3 times the length of the 7–11mm-long ovary. **Flowers** white with exposed parts tinged with light to dark purple, elongate tips of the segments soon drying out. Sepals 5–10mm long; petals similar to lateral sepals; lip flap 2–3mm long, reflexed; spurs 5–10mm long.

Distinguishable from *S. emarcidum* by the white to dark purple flower colour and lowermost leaves spreading near the ground.

| A | S | O | N | D | J | F | M | A | M | J | J |

Locally frequent in a wide variety of habitats, from 100–2,450m. Pollinated by moths and butterflies. Hybridizes with *S. acuminatum*. Flowers between late October and January.

Satyrium longicauda

Port Edward 19.10.2012

Port St Johns 21.10.2011

Maclear 18.12.2010

Sabie 24.01.2011

Ngome 10.01.2014

Slender to robust terrestrial 170–810mm tall; sterile shoot present. **Leaves** 1 or 2, on sterile shoot, egg-shaped to elliptic, 40–200mm long. **Inflorescence** lax, 15–60-flowered; bracts strongly deflexed, 1.4–1.8 times the length of the 6–15mm-long ovary. **Flowers** variable in colour, white to pink, veins usually with deeper colour, spurs and distal parts of sepals and petals tinged with deep pink. Sepals free almost to the base; lip with prominent, erect shortly hairy flap; spurs 15–46mm long.

Small forms have been recognized as var. *jacottetianum* but the large variation in this species makes the recognition of only a single variety difficult. Distinguishable from *S. neglectum* by the shortly hairy lip flap.

A S O N D J F M A M J J

Common in open grassland on moist and peaty soils, from sea level to 2,300m. Pollinated by hawkmoths. Flowers mostly between December and March, and as early as October near the coast.

Satyrium longicolle

Riversdale 04.10.2010   Krakeel 15.11.2010   Joubertina 20.10.2010

Mossel Bay 21.10.2011   Mossel Bay 21.10.2011

Slender to robust terrestrial 160–320mm tall. **Leaves** 2, pressed flat on the ground, broadly egg-shaped to almost circular, 20–110mm long. **Inflorescence** 10–26-flowered; bracts deflexed, 1.4–1.9 times the length of the 10–21mm-long ovary. **Flowers** pale pink to ivory with dark purple along midveins of petals and sepals and near lip entrance, anthers dark purple. Sepals and petals fused to the sides of the lip for a third of their length; lip tall and narrow, entrance 7–14mm tall, flap erect; spurs 19–30mm long.

A S O N D J F M A M J J

Rare to locally common on damp slopes, often in moist peaty soil, from near sea level to 750m. Flowers mainly between September and October, and occasionally up to December, after fire.

Satyrium lupulinum

Cape Peninsula 23.09.2007

Somerset West 14.10.2007

Paarl 26.09.2009

Hermanus 14.09.2006

Paarl 26.09.2009

Slender to fairly stout terrestrial 150–250mm tall. **Leaves** 1–3, spreading near the ground, egg-shaped, undersides often purple, 20–80mm long. **Inflorescence** rather dense, 12–20-flowered; bracts more than twice the length of the 7–12mm-long ovary, strongly deflexed. **Flowers** dull yellowish-green, usually tinged with dark purple. Sepals free almost to their bases; lip entrance 4–6mm tall, flap 2mm long, bent backward; spurs 15–18mm long.

A S O N D J F M A M J J

Occasional to locally frequent on moist, sandy slopes or flats, from 130–900m. Flowers between September and October, after fire.

Satyrium macrophyllum

Nottingham Road 06.03.2012

Loteni 28.02.2009

Nottingham Road 06.03.2012

Nottingham Road 04.03.2014

Loteni 25.02.2007

Robust terrestrial 250mm–1m tall. **Leaves** 2 or 3, basal, partly erect, egg- to spear-shaped, up to 400mm long. **Inflorescence** moderately lax, flowers held almost horizontally, 20–80-flowered; bracts fully deflexed, 2–4 times the length of the 8–13mm-long ovary. **Flowers** white to deep pink, tinged or streaked darker. Sepals 7–14mm long; fused to petals and lip for two thirds to three quarters of their length; lip much deeper than tall, entrance 3–5mm tall, flap 0.5mm long, bent backward; spurs 13–26mm long.

Some colour forms of this species can be confused with *S. cristatum* var. *cristatum*.

A S O N D J F M A M J J

Occasional on moist grassy slopes and flats, from 350–1,800m. Flowers between February and March.

Satyrium membranaceum

Mossel Bay 22.10.2010

Mossel Bay 05.11.2012

Port Elizabeth 30.09.2011

Mossel Bay 05.11.2012

Slender to robust terrestrial 230–480mm tall. **Leaves** 2, pressed flat on the ground, egg-shaped to elliptic, 20–120mm long; sheaths above dry and membranous. **Inflorescence** fairly dense, 8–22-flowered; bracts dry and membranous, gradually becoming deflexed, 1.3–1.9 times the length of the 9–18mm-long ovary. **Flowers** pale to deep pink, rarely almost white. Sepals and petals fused to lip for a quarter of their length; sepals 9–11mm long; petals fringed with thick hairs; lip 5–7mm high, 6–8.5mm deep, flap 3mm long, curled backward, fringed with thick hairs; spurs 21–27mm long.

Distinguishable from *S. jacottetiae* by the long, narrow rostellum, with enlarged, rounded apex presenting the sickle-shaped viscidia from notches on the side.

Occasional on grassy, open slopes, from near sea level to 700m. Flowers between September and November, stimulated by fire.

Satyrium microrrhynchum

Witsieshoek 08.02.2009

Sabie 30.01.2011

Naudesnek Pass 12.02.2008

Naudesnek Pass 07.02.2008

Slender terrestrial 150–280mm tall. **Leaves** 2 or 3, cauline, elliptic, 50–120mm long. **Inflorescence** dense, 12–56-flowered; bracts sharply deflexed, 1.5–3 times the length of the 5–8mm-long ovary. **Flowers** yellowish-green to dull white, thick-textured, with many short hairs on basal half of inner surfaces of petals and lip. Sepals 3–5mm long; petals slightly shorter; lip taller than deep, flap prominent, erect; spurs absent or as slight sacs at lip base.

A S O N D J F M A M J J

Rare on grassy and sometimes stony or moist slopes, stream banks and in seepages, from 1,600–3,300m. Pollinated by beetles and wasps. Flowers between January and February.

Satyrium muticum

Mossel Bay 23.08.2012

Mossel Bay 04.09.2010

Mossel Bay 04.09.2010

Mossel Bay 04.09.2010

Robust terrestrial 240–300mm tall. **Leaves** 1 or 2, pressed flat on the ground, broadly egg-shaped to circular, 30–90mm long. **Inflorescence** moderately dense, 3–12-flowered; bracts becoming fully deflexed at flowering time, 2.2–2.7 times the length of the 9–14mm-long ovary. **Flowers** pink, petals ivory with purple speckles along midvein, lip with dark pink speckles inside. Sepals and petals fused to the lip for a quarter of their length; sepals 14–16mm long; petals prominently papillose to torn, crisped; lip entrance 7mm tall, flap 4mm long, erect; spurs absent to rudimentary, up to 2mm long.

A S O N D J F M A M J J

Very rare on relatively dry to moist stony slopes, from sea level to 380m. Flowers between August and September.

Satyrium neglectum

Slender to robust terrestrial 380–810mm tall; sterile shoot present. **Leaves** usually 2 on sterile shoot, egg-shaped to oblong, 40–320mm long. **Inflorescence** dense, 17–80-flowered; bracts spear-shaped, 1.3–3.3 times the length of the 8–12mm-long ovary. **Flowers** pink to deep rose, sometimes yellowish. Sepals and petals fused to the lip for one third their length, curved in distal third; sepals elliptic, 7–10mm long; petals egg-shaped; lip approximately 3 times deeper than tall, entrance 2–9mm tall, lip flap 2mm long, without hairs; spurs 6–19mm long.

Satyrium neglectum subsp. *neglectum*

Luneburg 04.01.2012

Luneburg 10.01.2013

Luneburg 10.01.2013

Subsp. *neglectum* is distinguished as follows: entrance of lip 2–4mm tall; later flowering time.

Subsp. *neglectum* is widespread in the south-eastern parts of Africa, including South Africa, Swaziland and Lesotho.

Distinguishable from *S. longicauda* by the smooth lip flap.

Occasional and rather local in grassland, on moist slopes, from 1,200–1,600m. Flowers between January and March.

A S O N D J F M A M J J

Satyrium neglectum subsp. *woodii*

Eston 29.11.2012

Eston 29.11.2012

Eston 29.11.2012

Subsp. *woodii* is distinguished as follows: entrance of lip mostly 6–9mm tall; earlier flowering time.

Subsp. *woodii* is restricted to the Eastern Cape and the KwaZulu-Natal Midlands.

Distinguishable from *S. longicauda* by the smooth lip flap.

Rare and local in moist grassland, from 300–900m. Flowers between October and December.

A S O N D J F M A M J J

Satyrium odorum

Cape Peninsula 21.09.2008

Cape Peninsula 02.09.2010

Cape Peninsula 02.09.2010

Botrivier 27.08.2007

Robust terrestrial 200–440mm tall. **Leaves** 2–6, cauline, egg-shaped, 40–240mm long. **Inflorescence** moderately dense, 7–23-flowered; bracts strongly deflexed, 1.3–2.6 times the length of the 9–13mm-long ovary. **Flowers** pale green and yellowish, tinged with dull purple-brown on outer surfaces. Sepals deflexed; median sepal 3–7mm long; lateral sepals slightly longer; petals slightly more than half the length of the lateral sepals, less deflexed; lip entrance 4–6mm tall, flap 2mm long, bent backward; spurs 13–18mm long.

A	S	O	N	D	J	F	M	A	M	J	J

Common in partly shaded, humus-rich soils, near trees and bushes and especially near large boulders and rock ledges, from sea level to 800m. Flowers between August and October, stimulated by fire.

Satyrium outeniquense

Misgund 12.11.2011

Misgund 23.11.2011

Joubertina 15.11.2010

Slender terrestrial 140–280mm tall. **Leaves** 2–4, lowest flat on the ground, egg-shaped, 10–40mm long. **Inflorescence** lax, 6–15-flowered; bracts deflexed, 1.2–2.6 times the length of the 8–10mm-long ovary. **Flowers** pale greenish-yellow. Sepals and petals free almost to the base; sepals 6–7mm long; petals slightly smaller, with somewhat crisped margins; lip deeper than tall, flap 2mm long, bent backward; spurs 12–16mm long.

Distinguishable from *S. stenopetalum* by the deflexed bracts.

| A | S | **O** | **N** | D | J | F | M | A | M | J | J |

Rare to locally common in dry to moist fynbos, on sandy, stony soils, from 160–1,000m. Flowers between October and November, after fire.

Satyrium pallens

Joubertina 02.10.2010

Joubertina 02.10.2010

Ceres 24.09.2010

Avontuur 20.10.2013

Slender to robust terrestrial up to 500mm tall. **Leaves** 2, pressed flat on the ground, egg-shaped to elliptic, up to 100mm long. **Inflorescence** lax to dense, 10–25-flowered; bracts deflexed. **Flowers** white or cream, or tinged with pale pink. Sepals fused to petals only right at the base, 10–15mm long; petals similar; lip taller than deep, entrance 10–12mm tall, flap 2–3mm long, erect; spurs 19–22mm long.

Distinguishable from *S. erectum* by the longer spurs and the absence of a pungent odour.

Occasional and local in dry fynbos and karroid vegetation, on rocky south-facing slopes, from 500–1,400m. Pollinated by hawkmoths. Hybridizes with *S. coriifolium* and *S. erectum*. Flowers between September and October.

Satyrium parviflorum

Witsieshoek 20.02.2009

Naudesnek Pass 23.02.2007

Mossel Bay 27.10.2010

Bathurst 23.10.2010

Mossel Bay 27.10.2010

Slender to robust terrestrial 210–750mm tall. **Leaves** variable, from absent or partly developed on a sterile shoot to 2–4 at base of flowering shoot, elliptic to broadly egg-shaped, spreading, 10–330mm long. **Inflorescence** lax to somewhat dense, 20–100-flowered; bracts deflexed, 2.5 times the length of the 4–10mm-long ovary. **Flowers** yellowish-green to olive or dark maroon, petals drying brown soon after the flowers open. Sepals and petals fused to the lip for a quarter of their length; median sepal 2–4mm long; lateral sepals longer; lip more or less egg-shaped, entrance very small; spurs often curved away from ovary at first, 5–15mm long. Highly variable in all aspects.

| A | S | O | N | D | J | F | M | A | M | J | J |

Widespread, but never common, in a wide variety of habitats ranging from fynbos to marshy, moist or dry places in grassland and forest, from near sea level to 2,600m. Flowers between September and March.

Satyrium princeps

Port Elizabeth 30.09.2011

Sedgefield 04.11.2007

Sedgefield 04.11.2007

Robust terrestrial 340–690mm tall. **Leaves** 2, pressed flat on the ground, very broadly egg-shaped to elliptic; sheaths 7–9, dry and membranous at flowering time. **Inflorescence** dense, 13–33-flowered; bracts gradually becoming deflexed, 1.4–2.8 times the length of the 8–17mm-long ovary. **Flowers** rose-pink to carmine-red. Sepals and petals fused to the lip for one fifth of their length; median sepal 11–14mm long; lip entrance 7–8mm tall, flap thickly hairy and crisped, partly reflexed; spurs 16–20mm long.

A S O N D J F M A M J J

Occasional in open places, between bushes on dunes close to the shoreline, from sea level to 150m. Pollinated by sunbirds. Flowers between late September and early November.

Satyrium pulchrum

Knersvlakte 30.08.1996

Knersvlakte 18.08.2010

Knersvlakte 18.08.2010

Slender to robust lithophyte up to 400mm tall. **Leaves** 2, pressed flat on the ground, egg-shaped to elliptic, up to 110mm long. **Inflorescence** lax to dense, up to 20-flowered; bracts deflexed. **Flowers** pink, sepals and petals with white patch at the base. Sepals fused to petals in lower third; petals similar to sepals; lip taller than deep, entrance 9–10mm tall, flap 1–2mm long, erect; spurs 18–22mm long.

Distinguishable from *S. erectum* by the plain pink flowers usually without spots or stripes, longer spurs and the lip that is deeper than tall.

| A | S | O | N | D | J | F | M | A | M | J | J |

Rare on granite inselbergs, growing in rock clefts, from 500–600m. Flowers between August and September.

Satyrium pumilum

Op-die-Berg 09.10.2007

Ceres 24.09.2010

Ceres 24.09.2010

Rosette-like terrestrial 10–30mm tall. **Leaves** 3–5, lowest egg-shaped to tapering, spreading near the ground, 20–55mm long. **Inflorescence** dense, with 2–4 relatively large flowers; bracts 3–8 times the length of the 5–10mm-long ovary. **Flowers** dull green outside, lip dull greenish-yellow inside marked with transverse bars of dark maroon. Sepals and petals fused in basal three quarters to two thirds, forming a broadly egg-shaped lamina 8–11mm long; lateral sepals strongly curved; lip 9–12mm long, flap erect; spurs pouch-shaped, 2–3mm long.

A S O N D J F M A M J J

Rare to locally abundant in lower-rainfall areas, in clay, sandy soils or on rock ledges with a slight seepage, from 200–1,420m. Pollinated by flies. Flowers between September and October, stimulated by fire.

Satyrium pygmaeum

Ceres 09.10.2009

Grabouw 24.10.2009

Ceres 07.11.2008

Ceres 10.10.2009

Slender terrestrial 90–160mm tall. **Leaves** 1–3, lowest spreading near the ground, upper partly erect, egg-shaped, 10–40mm long. **Inflorescence** moderately dense, 15–30-flowered; bracts deflexed at flowering time, 1.3–3.4 times the length of the 5–7mm-long ovary. **Flowers** dull yellowish-green tinged with pale to dark purple. Sepals fused to petals and base of the lip for one quarter of their length, then abruptly deflexed with the petals, 3–4mm long; median sepal much narrower than lateral sepals; lip an almost flat to slightly hooded blade, flap absent; spurs 6–10mm long, standing away from ovary.

A S O N D J F M A M J J

Rare to locally abundant on wet, mossy ledges on sheltered slopes and cliffs, mainly from 500–1,000m. Flowers between October and November, after fire.

Satyrium retusum

Franschhoek 08.11.2007

Grabouw 08.11.2007

Grabouw 27.11.2009

Grabouw 10.11.2009

Slender to fairly robust terrestrial 80–150mm tall. **Leaves** 1–5, cauline, spreading, broadly egg-shaped, 7–60mm long. **Inflorescence** dense, many-flowered; bracts pointing upward, 2–4 times the length of the 4–7mm-long ovary. **Flowers** greenish to yellowish-white, lip with deep red lines inside. Sepals 1–4mm long; median sepal projecting forward, margins deflexed; lateral sepals spreading widely from side of lip entrance; petals projecting forward, about as long as median sepal; lip taller than deep; flap absent or weak; spurs variable, pouch- to club-shaped and curved, 0.3–2.7mm long.

Distinguishable from *S. bicallosum* by the lip apex that is not folded downward in front of the column.

A S O N D J F M A M J J

Rare in acid, peaty to sandy, moist soils, from near sea level to 1,700m. Flowers between October and December, after fire.

Satyrium rhodanthum

Ixopo 17.11.2011

Ixopo 17.11.2011

Ixopo 17.11.2011

Fairly robust terrestrial 250–400mm tall; sterile shoot present. **Leaves** 2, on sterile shoot, developing after flowering; fertile stem almost entirely covered by sheaths. **Inflorescence** dense, 10–25-flowered; bracts strongly deflexed, lower ones longer than flowers, upper ones shorter. **Flowers** deep pink to carmine-red. Sepals and petals free, recurved, similar, 13mm long; lip deeper than tall; flap erect; spurs thread-like, deflexed, twice as long as ovary.

A S O N D J F M A M J J

Very rare in damp grassland, from 700–1,100m. Pollinated by sunbirds. Flowers between November and December.

Satyrium rhynchanthum

Grabouw 07.12.2009

Franschhoek 29.11.2007

Franschhoek 14.12.2007

Grabouw 12.12.2009

Slender terrestrial 160–400mm tall. **Leaves** 2–6, cauline, linear to spear-shaped, 30–90mm long. **Inflorescence** rather densely 10–60-flowered; bracts fully deflexed, mostly shorter than the 7–9mm-long, slender ovary. **Flowers** white to pale pink with dark purple spots inside lip and spurs and on petal bases, spurs with outer surface pale green. Sepals 5–7mm long; petals similar to sepals; lip galea very shallow, flap elongate, ascending; spurs facing backward, 5–6mm long.

| A | S | O | N | D | J | F | M | A | M | J | J |

Rather rare, but sometimes locally abundant in fynbos, marshes or on moist slopes, from 20–1,000m. Pollinated by oil-collecting bees. Flowers between November and December, after fire.

Satyrium rupestre

George 12.12.2010

George 02.11.2012

George 20.10.2013

George 25.11.2013

George 25.11.2013

Slender terrestrial 190–300mm tall. **Leaves** 2–4, borne on lower part of the stem, egg-shaped, 40–150mm long. **Inflorescence** rather dense, 6–45-flowered; bracts deflexed, 1.7–2.7 times the length of the 3–7mm-long ovary. **Flowers** white with pale green tinges along midveins. Sepals, petals and lip fused in basal quarter; median sepal reflexed earlier than the rest; lateral sepals slightly larger; lip flap minute, bent backward; spurs 2–4mm long.

A S O N D J F M A M J J

Rare in damp, shaded places in forests and on damp, mossy rock faces, from 60–1,500m. Flowers between late October and December.

Satyrium situsanguinum

Grabouw 24.10.2009

Citrusdal 18.10.2009

Citrusdal 16.10.2009

Citrusdal 11.10.2007

Grabouw 03.12.2009

Slender terrestrial 160–420mm tall. **Leaves** 2, basal, flat on the ground, broadly egg-shaped, 20–68mm long; with cup-like leaf sheaths higher up.
Inflorescence lax, 2–31-flowered; bracts deflexed. **Flowers** white, occasionally cream-yellow, sepals and petals sometimes with light pink tinges. Sepals and petals fused for about one fifth to a quarter of their length; lip deeper than tall, entrance 2.5–4mm tall, flap prominent, erect to bent backward; spurs 15–22mm long.

Distinguishable from several other similar *Satyrium* species by the combination of the following characters: cup-shaped leaf sheaths, white flowers with spurs longer than the ovary and a rostellum with lateral viscidia.

| A | S | O | N | D | J | F | M | A | M | J | J |

Rare on south-facing rock ledges, from 1,000–1,400m. Flowers between October and early December, stimulated by fire.

Satyrium sphaerocarpum

Ngeli 30.01.2010

Paddock 30.01.2010

Naudesnek Pass 16.02.2008

Ngeli 17.01.2014

Slender to robust terrestrial 220–400mm tall. **Leaves** 2, rarely 1, spreading near the ground, oblong to egg-shaped, 30–170mm long. **Inflorescence** dense, 8–27-flowered; bracts partly to fully deflexed, 2.5 times the length of the 7–9mm-long ovary. **Flowers** white to cream, inner surfaces variously marked with brownish-red to red-purple. Sepals 13–18mm long, fused to petals and lower side of lip for about a quarter to a third of their length; lip taller than deep, entrance 7–10mm tall, flap 1mm long, bent backward; spurs slightly longer to slightly shorter than ovary, 8–15mm long.

A S O N D J F M A M J J

Rare to locally frequent in moist grassy uplands and coastal sandy flats, from close to sea level to 2,000m. Pollinated by bees. Flowers between October and March.

Satyrium stenopetalum

Slender to robust terrestrial 220–400mm tall. **Leaves** 3–7, cauline but concentrated in lower part of the stem, elliptic, 20–80mm long. **Inflorescence** rather lax, 10–27-flowered; bracts erect, 0.9–2 times the length of the 12–21mm-long ovary. **Flowers** white, faintly tinged with cream, with traces of pink. Sepals free almost to the base; sepals 5–13mm long; petals similar but more recurved; lip flap conspicuous, 4mm long, bent backward; spurs 9–30mm long.

Satyrium stenopetalum subsp. brevicalcaratum

Villiersdorp 28.10.2008

Kleinmond 14.10.2011

Pringle Bay 02.11.2008

Ceres 10.10.2007

Wolseley 21.10.2006

Subsp. *brevicalcaratum* is distinguished as follows: spurs 9–14mm long; lateral sepals 5–10mm long.

Subsp. *brevicalcaratum* is restricted to the Western Cape, from Clanwilliam to Bredasdorp.

Rather rare to locally common on moist sandy, marshy flats and along rocky perennial streams, from near sea level to 1,100m. Pollinated by moths. Flowers between September and December but mostly in November, stimulated by fire.

A S O N D J F M A M J J

Satyrium stenopetalum subsp. *stenopetalum*

Storms River Mouth 13.11.2010

Storms River Mouth 13.12.2007

Louterwater 14.11.2010

Subsp. *stenopetalum* is distinguished as follows: spurs 19–30mm long; lateral sepals 7–13mm long.

Subsp. *stenopetalum* is restricted to the southern part of the Cape Floral Region between Swellendam and Humansdorp.

Distinguishable from *S. outeniquense* by the erect bracts.

Occasional in coastal forelands, in open stony, marshy or dry places among grass and fynbos, from 50–700m. Flowers between September and December but mostly in November, stimulated by fire.

| A | S | O | N | D | J | F | M | A | M | J | J |

Satyrium striatum

Porterville 23.09.2009

Op-die-Berg 13.10.2008

Op-die-Berg 18.09.2013

Slender terrestrial 70–90mm tall, sticky on most surfaces except flowers, usually retaining soil particles on them. **Leaves** 1 or 2, basal, broadly egg-shaped, up to 21mm long, with a gradual transition to conical leaf sheaths higher up. **Inflorescence** dense, 1–8-flowered; bracts erect, 1–2 times the length of the 6–7mm-long ovary. **Flowers** dull yellow to greenish-white, with dark purple striped markings. Sepals 4–6mm long; petals narrower; lip curved but not hooded over the column, 5–8mm long, flap absent; spurs rounded to almost conical, 2mm long.

A S O N D J F M A M J J

Very rare on moist flats and slopes in coarse, sandstone-derived, often stony soils, from 100–1,100m. Flowers between September and October, stimulated by fire.

Satyrium trinerve

Graskop 01.01.2011

Umtamvuna 01.02.2009

Graskop 11.01.2010

Barberton 06.01.2014

Umtamvuna 01.02.2009

Slender to moderately robust terrestrial 350–550mm tall. **Leaves** 2–5, cauline, spear-shaped, nearly erect, 60–250mm long. **Inflorescence** dense, 15–60-flowered; bracts conspicuous, white, spreading at 90° from the stem, 3–8 times the length of the 5–8mm-long ovary. **Flowers** white, median sepal and petals lemon-yellow. Median sepal to 4mm long; lateral sepals slightly longer; petals fused to sepals in basal third; lip slightly taller than deep, flap prominent, erect; spurs pouch-shaped, 1–5mm long.

A S O N D J F M A M J J

Locally common in black, wet soils, from near sea level to 2,200m. Pollinated mainly by beetles. Flowers between October and February.

Schizochilus angustifolius

Sentinel 15.01.2008

Sentinel 02.01.2009

Sentinel 21.01.2014

Sentinel 02.01.2009

Sentinel 23.01.2014

Slender terrestrial up to 200mm tall. **Leaves** 7–14, with about two thirds in a basal cluster, linear to narrowly spear-shaped, 30–60mm long; **cauline leaves** smaller. **Inflorescence** dense, nodding, up to 40mm long, 10–30-flowered. **Flowers** white, lip yellow; 2–3mm in diameter. Sepals 4–5mm long; petals 2mm long; lip 3.5–5mm long; lip midlobe 1.5mm long; lip side lobes 0.8mm long; disc without calli; spur 1mm long.

Distinguishable from *S. flexuosus* by the shorter spur.

A S O N D J F M A M J J

Localized on pebbly or rocky ridges, in alpine and subalpine grassland, from 2,100–3,000m. Pollinated by flies. Flowers between January and February.

Schizochilus bulbinella

Ramatseliso 03.02.2008

Ramatseliso 03.02.2008

Ramatseliso 07.01.2009

Ramatseliso 03.02.2008

Ramatseliso 07.01.2009

Slender terrestrial up to 250mm tall. **Leaves** 10–30, with 4–20 in a basal cluster, very narrowly spear-shaped, 30–100mm long; **cauline leaves** smaller. **Inflorescence** dense, 20–40mm long, 20–70-flowered. **Flowers** bright yellow, 3mm in diameter. Sepals 2.5–3.5mm long; petals 1.5–2mm long; lip 3–4 × 1–2mm; lip midlobe 1mm long; lip side lobes 0.3mm long or less; disc without callus; spur 0.3mm long.

Distinguishable from *S. zeyheri* by the shorter spur and the rocky high-altitude grassland habitat.

| A | S | O | N | D | J | F | M | A | M | J | J |

Frequent on shallow soil over rock, in montane and alpine grassland, from 1,500–2,500m. Pollinated by flies. Hybridizes with *S. flexuosus*. Flowers between January and February.

Schizochilus cecilii

Slender terrestrial up to 300mm tall. **Leaves** 6–15, with 3–6 in a basal cluster, linear to narrowly spear-shaped, up to 100mm long; **cauline leaves** smaller. **Inflorescence** dense, nodding, up to 70mm long, 12–40-flowered. **Flowers** bright yellow or white, 2–3mm in diameter. Sepals 2–4mm long; petals 1.3–3mm long; lip 3–3.6 × 1.5–1.8mm; lip midlobe approximately 1mm long; lip side lobes less than half as long as midlobe; disc with 3 fleshy calli; spur cylindrical, 0.1–1mm long.

Schizochilus cecilii subsp. culveri

Verloren Vallei 14.01.2010

Verloren Vallei 31.01.2014

Verloren Vallei 08.01.2013

Verloren Vallei 31.01.2014

Subsp. *culveri* is distinguished as follows: flowers yellow; sepals 2.5–3.5mm long; spur 0.8–1mm long.

Subsp. *culveri* is restricted to Mpumalanga and Swaziland.

Distinguishable from *S. zeyheri* by the smaller flowers and shorter spur.

Relatively rare on rock ledges in grassland, from 1,000–2,200m. Flowers between December and January.

A S O N D J F M A M J J

Schizochilus cecilii subsp. *transvaalensis*

Mokobulaan 02.01.2011

Sabie 25.01.2014

Sabie 01.01.2011

Sabie 25.01.2014

Sabie 30.01.2011

Subsp. *transvaalensis* is distinguished as follows: flowers white or cream or, very rarely, yellow; median sepal 2.8–3.8mm long; lateral sepals 3.5–4.5mm long; spur 0.2–0.6mm long.

Subsp. *transvaalensis* is restricted to Mpumalanga and Limpopo.

Common in well-drained, often pebbly soil in grasslands, from 1,000–2,200m. Flowers between December and February.

| A | S | O | N | D | J | F | M | A | M | J | J |

Schizochilus crenulatus

Graskop 12.01.2010

Graskop 08.01.2011

Graskop 08.01.2011

Graskop 04.01.2011

Slender terrestrial up to 350mm tall. **Leaves** 5 or 6, with 2–4 in a basal cluster, linear, 70 × 4mm; **cauline leaves** smaller. **Inflorescence** lax, nodding, 30–150mm long, 6–25-flowered. **Flowers** white with veins mauve, 5mm in diameter. Sepals 5–7mm long; petals 2.5–3mm long; lip 6–7 × 3–4mm; lip midlobe 2mm long; lip side lobes 1mm long, usually notched with small, regular teeth; disc with a ridge or filamentous crests; spur more or less club-shaped, 2–3mm long.

Distinguishable from *S. lilacinus* by the linear leaves, 4–7mm wide, and later flowering time.

| A | S | O | N | D | J | F | M | A | M | J | J |

Very localized in seepages, in running water on rock sheets or in wet grassland, from 1,500–1,750m. Flowers between December and February.

Schizochilus flexuosus

Garden Castle 26.01.2010

Naudesnek Pass 07.02.2008

Luneburg 08.01.2014

Garden Castle 03.02.2007

Slender terrestrial up to 250mm tall. **Leaves** 10–30, with 6–20 in a basal cluster, linear to spear-shaped, 20–50 × 2–8mm; **cauline leaves** smaller. **Inflorescence** dense to almost head-like, nodding, 20–40mm long, 5–20-flowered. **Flowers** with sepals and petals white, lip yellow; 5mm in diameter. Sepals 5–9mm long; petals 2.5–4.5mm long; lip 4–8 × 3–4mm; lip midlobe 2mm long; lip side lobes 1mm long; disc calli rudimentary or absent; spur more or less cylindrical to more or less club-shaped, 3–4mm long.

   Distinguishable from *S. angustifolius* by the longer spur.

A S O N D J F M A M J J

Localized but often fairly common in grassland, from 1,500–2,500m. Hybridizes with *S. bulbinella* and *S. zeyheri*. Pollinated by bees. Flowers between January and February.

Schizochilus gerrardii

Ngome 05.12.2011

Ngome 05.12.2011

Ngome 09.12.2012

Ngome 10.01.2014

Slender terrestrial up to 300mm tall. **Leaves** up to 25, with 6–10 in a basal cluster, linear, up to 110 × 10mm; **cauline leaves** smaller. **Inflorescence** dense, nodding, up to 50 × 15mm, 50-flowered. **Flowers** white suffused with pink or mauve, lip with a green spot; 5–8mm in diameter. Sepals 5–8mm long; petals 2.5–3mm long; lip 4–6 × 2mm; lip midlobe 1mm long; lip side lobes 0.1–0.2mm long; disc with 3 calli; spur cylindrical, 1mm long.

A S O N D J F M A M J J

Very localized along margins of rock outcrops in grassland, at around 1,200m. Flowers between December and January.

Schizochilus lilacinus

Sabie 28.10.2010

Sabie 18.11.2013

Sabie 05.11.2011

Sabie 18.11.2013

Slender terrestrial up to 350mm tall. **Leaves** 3 or 4, with 2 or 3 near the base, narrowly elliptical, 40–100 × 17–25mm long; **cauline leaves** smaller. **Inflorescence** lax to semi-dense, nodding, up to 100mm long, 3–25-flowered. **Flowers** white or white with pink veins, 10mm in diameter. Sepals 8–14mm long; petals 3mm long; lip 6–8 × 2.5–3.5mm, obscurely 3-lobed; lip midlobe 2.5–3.5mm long; lip side lobes 0.2mm long; disc with a small tuft of crests; spur straight, deeply divided in two, 1.5–3mm long.

Distinguishable from *S. crenulatus* by the elliptic leaves, 17–25mm wide, and earlier flowering time.

A S **O N D** J F M A M J J

Rare and very localized on ledges of very steep rock faces, in montane grassland, from 1,600–2,300m. Flowers between October and November.

Schizochilus zeyheri

Engcobo 09.01.2009

Sabie 26.01.2011

Luneburg 22.01.2011

Vryheid 09.01.2014

Vryheid 09.01.2014

Slender terrestrial up to 600mm tall. **Leaves** 5–18, with 3–10 near the base or spread over lower half of the stem, linear to very narrowly elliptical, 40–140 × 3–10mm; **cauline leaves** smaller. **Inflorescence** lax or rarely dense, nodding, up to 100mm long, up to 30-flowered. **Flowers** bright yellow, lip a richer yellow than sepals; 4–6mm in diameter. Sepals 4.5–11mm long; petals 2.5–6mm long; lip 6–12 × 4–6mm; lip midlobe up to 2mm long; lip side lobes 1mm long; disc with 3 fleshy calli; spur cylindrical, 2–6mm long.

Distinguishable from *S. bulbinella* by the longer spur and its wet, lower-altitude grassland habitat, and from *S. cecilii* subsp. *culveri* by the larger flowers and longer spur.

| A | S | O | N | D | J | F | M | A | M | J | J |

Widespread and relatively common in wet grasslands, from sea level to 2,000m. Pollinated by flies. Hybridizes with *S. flexuosus*. Flowers between December and February.

Stenoglottis fimbriata

Giant's Castle 05.01.2009

Giant's Castle 04.02.2007

Giant's Castle 26.02.2008

Giant's Castle 19.01.2014

Kei Mouth 28.11.2013

Giant's Castle 19.01.2014

Slender lithophyte, terrestrial or epiphyte, up to 400mm tall. **Leaves** 6–10, in a basal rosette, strap-shaped to narrowly spear-shaped, spotted, margins strongly wavy, 25–150 × 5–16mm. **Inflorescence** lax, 5–30-flowered. **Flowers** lilac with darker spots. Sepals similar, 3–8mm long; petals shorter than sepals, enclosing the column; lip 3-lobed, 6–15mm long; lip midlobe usually longer and narrower than side lobes; spur absent. A very variable species, both in size and coloration of all parts.

Distinguishable from *S. zambesiaca* by the lip midlobe that is longer and narrower than the side lobes, and from *S. longifolia* by the 3-lobed lip, shorter growth form and fewer flowers.

A S O N D J F M A M J J

Fairly common on mossy substrates in forests, on rocks, from near sea level to 2,200m. Flowers between January and March.

Stenoglottis longifolia

Eshowe 06.03.2013

Eshowe 20.03.2013

Eshowe 20.03.2013

Fairly robust lithophyte or terrestrial, up to 600mm tall. **Leaves** 6–10, in a basal rosette, strap-shaped to narrowly elliptic, uniformly green, margins slightly wavy, 60–150 × 12–25mm. **Inflorescence** lax to dense, up to 80-flowered. **Flowers** lilac to white, with darker spots on all parts. Sepals similar, 7–10mm long; petals shorter than sepals, enclosing the column; lip 5-lobed; lip midlobe longer than the two pairs of side lobes; lip side lobes sometimes united into a single, torn lobe; spur absent.

Distinguishable from *S. fimbriata* by the 5-lobed lip, taller growth form and flowers that are more numerous.

| A | S | O | N | D | J | F | M | A | M | J | J |

Rare and localized, from 300–700m. Flowers between February and May.

Stenoglottis woodii

Peacevale 05.03.2013

Locality unknown (cultivated plant) 28.01.2013

Locality unknown (cultivated plant) 28.01.2013

Slender terrestrial or lithophyte, up to 200mm tall. **Leaves** 5–20, in a basal rosette, narrowly elliptic to egg-shaped, uniformly coloured, 30–80 × 9–16mm. **Inflorescence** lax, 4–20-flowered. **Flowers** white to pink, lip often with purple spots at the base. Sepals similar, 4–5mm long; petals erect, 3.5mm long; lip 3-lobed, 5–9mm long; lip side lobes larger than midlobe; spur 2–3mm long.

A S O N D J F M A M J J

Occasional in crevices in sandstone cliffs, from near sea level to 1,500m. Flowers between December and March.

Stenoglottis zambesiaca

Ngome 24.02.2011

Lydenburg 14.02.2010

Tzaneen 09.03.2013

Graskop 07.02.2012

Very slender lithophyte or rarely an epiphyte, up to 350mm tall. **Leaves** 6–12, in a basal rosette, strap- to narrowly spear-shaped, uniformly coloured or sometimes spotted, slightly wavy, up to 120 × 8–20mm. **Inflorescence** lax, 5–40-flowered. **Flowers** pale lilac or pink, lip darker-spotted. Sepals similar; median sepal 3–5.5mm long; lateral sepals slightly longer; petals enclosing the column, 4–5mm long; lip 3-lobed, 5–12mm long; lip side lobes equalling or shorter and wider than midlobe; spur absent.

Distinguishable from *S. fimbriata* by the lip midlobe that is as long as or slightly longer than the blunt, rounded or squarely cut off side lobes.

| A | S | O | N | D | J | F | M | A | M | J | J |

Rare on rocks and tree trunks, from 1,300–2,150m. Flowers between February and April.

Tridactyle bicaudata

Slender to robust epiphyte or lithophyte; stems simple, stout, usually erect, up to 350mm long, 4–6mm in diameter; roots separated from leaves by a length of bare stem. **Leaves** in 2 opposite rows, leathery, linear to strap-shaped, rounded, unequally bilobed, 30–120 × 7–12mm. **Inflorescences** dense, appearing from the axils of the leaves; 30–80mm long, 6–20-flowered. **Flowers** facing upward, pale yellowish-brown, 10–14mm in diameter. Sepals spreading; median sepal narrowly oblong, 3.5–6 × 2–2.5mm; lateral sepals egg-shaped, 4–6 × 2.5–3mm; petals strap-shaped, 3.5–4.5 × 1–1.2mm; lip 3-lobed; lip side lobes frilled, usually longer than midlobe; spur 9–13mm long.

Tridactyle bicaudata subsp. *bicaudata*

Sedgefield 07.12.2008

Nature's Valley 14.12.2010

Sedgefield 05.12.2008

Sedgefield 05.12.2008

Sedgefield 27.11.2013

Subsp. *bicaudata* is distinguished as follows: growth form epiphytic; leaves flexuose, 60–120mm long.

Subsp. *bicaudata* is widespread in Africa.

Fairly common in dry forests, from near sea level to 1,400m. Flowers between October and February.

A S O N D J F M A M J J

Tridactyle bicaudata subsp. *rupestris*

Umtamvuna 18.12.2011

Umtamvuna 11.12.2011

Umtamvuna 11.12.2011

Subsp. *rupestris* is distinguished as follows: growth form lithophytic; leaves stiff, curved, 30–60mm long.

Subsp. *rupestris* is restricted to the South African coast from Port Elizabeth to southern KwaZulu-Natal.

Restricted to the coast, in full sun, from near sea level to 600m. Flowers between October and February.

A	S	O	N	D	J	F	M	A	M	J	J

Tridactyle gentilii

KZN north coast (cultivated plant) 28.01.2013

KZN north coast (cultivated plant) 28.01.2013

KZN north coast (cultivated plant) 12.02.2013

KZN north coast (cultivated plant) 12.02.2013

Slender epiphyte; stems reed-like, up to 400mm long, 7mm in diameter. **Leaves** linear to strap-shaped, unequally bilobed, 150 × 10mm. **Inflorescences** fairly dense, borne below the leaves, 50–70mm long, up to 8-flowered. **Flowers** facing upward, white, 16–20mm in diameter. Sepals spreading, reflexed apically; median sepal 7 × 4mm; lateral sepals 8.5 × 5mm; petals strap- to spear-shaped, spreading, reflexed apically, 7.5 × 2mm; lip deeply 3-lobed; lip side lobes frilled, usually longer than midlobe; spur 40–80mm long.

| A | S | O | N | D | J | F | M | A | M | J | J |

Very rare on trees in riverine bush, near sea level. Flowers between February and April.

Tridactyle tricuspis

Sabie 08.04.2013

Lydenburg 22.03.2010

Lydenburg 22.03.2010

Lydenburg 22.03.2010

Slender to robust epiphyte; stems erect, up to 200mm long, but usually much shorter, 6mm in diameter; roots produced along length of stem to base of leaves. **Leaves** closely overlapping, in 2 rows, linear to strap-shaped, unequally bilobed, 50–160 × 5–12mm. **Inflorescences** fairly dense, borne below the leaves, 40–120mm long, 10–25-flowered. **Flowers** greenish to cream, 10–11mm in diameter. Sepals strap-shaped, reflexed, 6–9 × 2mm; median sepal slightly narrower than lateral sepals; petals strap-shaped, reflexed, 5–6 × 1.5mm; lip 3-lobed; lip side lobes diverging, 7–9 × 2.5mm; spur 12–14mm long.

| A | S | O | N | D | J | F | M | A | M | J | J |

Common epiphyte growing high up in temperate forests, from 800–1,700m. Flowers between February and April.

Tridactyle tridentata

Umtamvuna 16.11.2011

Eshowe 29.06.2013

Eshowe 16.06.2012

Umtamvuna 11.11.2012

Slender epiphyte or lithophyte; stems up to 300mm long, branching, tangled, 3mm in diameter; roots restricted to the very base. **Leaves** widely spaced, in 2 rows, circular in cross section, grooved, stiff and often fleshy, 60–100 × 2–4mm. **Inflorescences** borne below the leaves, shorter than 5mm, 2–5-flowered. **Flowers** pale green to cream, 5–8mm in diameter. Sepals spreading, 3 × 2mm; petals strap-shaped, 2.5 × 1mm; lip 3-lobed, 2–3mm long; lip side lobes tooth-like, shorter than midlobe; spur 7–8mm long.

A S O N D J F M A M J J

Occasional in lowland forest, often near forest margins, from near sea level to 300m. Flowers between November and December.

Vanilla roscheri

KZN north coast 07.02.2014

KZN north coast 07.02.2014

KZN north coast 10.01.2014

KZN north coast 22.03.2013

Robust liana of indeterminate length; stem greenish-brown, cylindrical, succulent, up to 20mm in diameter, with a channel on either side and short roots arising from nodes. **Leaves** rudimentary, membranous, brown not green, up to 30mm long. **Inflorescences** terminal or axillary, up to 35-flowered. **Flowers** white, lip white with a pink throat. Sepals and petals spear-shaped to oblong, up to 80mm long; lip entire, up to 80 × 45mm, forming a funnel with the column, lip lamina with an irregular crest.

A S O N D J F M A M J J

Very localized in coastal bushland, mangrove swamps and open evergreen scrub, near sea level. Flowers between January and February.

Ypsilopus erectus

Syn: *Ypsilopus longifolius* subsp. *erectus*

Umtamvuna 12.04.2013

Umtamvuna 07.05.2013

Umtamvuna 12.04.2013

Slender epiphyte or lithophyte, up to 150mm tall; stems 20–30mm tall, erect; roots borne from the base of the stem. **Leaves** borne in an apical fan, linear to strap-shaped, up to 150 × 6.5mm. **Inflorescences** lax, several, arising below the leaves, up to 120mm long, 2–12-flowered. **Flowers** in 2 ranks, white. Sepals equal, reflexed, 4.5 × 2mm; petals like sepals but slightly smaller; lip shallowly 3-lobed, 4.5 × 3mm; spur curved, 45mm long.

A S O N D J F M A M J J

Localized near the coast, on rocks and trees in forested gorges, from 150–450m. Flowers between April and May.

Zeuxine africana

Durban 28.07.2013

Durban 28.07.2013

Durban 28.07.2013

Slender terrestrial up to 200mm tall. **Leaves** numerous, overlapping, erect, linear to strap-shaped, up to 60 × 5mm, grading into the bracts. **Inflorescence** cylindrical, many-flowered; bracts overtopping the flowers. **Flowers** white, lip yellow; 3mm in diameter. Sepals 1-veined, 2.5–3 × 1–1.5mm; petals 2.8 × 1mm; lip 2-veined, fiddle-shaped, basally sac-like with two calli, margins bending inward and somewhat uneven, 2.2mm long.

A S O N D J F M A M J J

Very rare in beach sand in the shade of *Imperata cylindrica*, near sea level. Flowers between July and August.

NATURAL HYBRIDS

More than 20 putative natural hybrids among South African orchid species have been reported, and these are mentioned on the relevant species pages in this guide. Natural hybrids can be identified by their morphology and coloration, which are typically intermediate between two parent species that grow together in the same habitat and which overlap in their flowering time. As a general rule, natural hybrids are formed only among species belonging to the same genus. Some previously reported 'hybrids' have, upon further investigation, turned out to be actual species. An example of this is *Satyrium rhodanthum* which was once erroneously considered to be a hybrid between *S. longicauda* and *S. neglectum* subsp. *woodii*. Most natural hybrids are extremely rare. Here we illustrate six of the more commonly encountered natural orchid hybrids in South Africa.

Satyrium erectum × Satyrium humile
Op-die-Berg 14.10.2009

Satyrium acuminatum × Satyrium ligulatum
Tsitsikamma 05.11.2007

Satyrium erectum × Satyrium bicorne
Vanrhynsdorp 11.09.2014

Satyrium erectum × Satyrium coriifolium
Ceres 13.09.2007

Ceratandra grandiflora × Ceratandra globosa
George 04.12.2008

Disa atricapilla × Disa bivalvata
Ceres 06.12.2008

IDENTIFICATION KEY TO THE SOUTH AFRICAN GENERA

1   a. Pollinia in pouch-like structures without a removable anther-cap; plants almost always terrestrial; underground organs almost always with root tubers, rarely clusters of thick roots with furry sheaths or bundles of thick fleshy roots (in the latter case the leaves in a basal rosette) 2
    b. Pollinia at the apex of the column under a removable anther-cap; if anther-cap insignificant or absent, underground organs not as above; plants terrestrial, epiphytic or lithophytic, never with root tubers 34

2   a. Column with a significant stalk which is longer than wide; anther pendent 3
    b. Column without such a stalk 4

3   a. Lip hooded, usually with 2 spurs or sacs at the base .... *Satyrium*
    b. Lip flat, without a spur; perianth lobes similar .... *Pachites*

4   a. Lip or its side lobes with a fringed or deeply dissected margin 5
    b. Lip not with a fringed margin; if lobed then with fewer than 6 lobes 10

5   a. Median sepal spurred .... *Disa*
    b. Median sepal not spurred 6

6   a. Lip with large appendage .... *Pterygodium*
    b. Lip without an appendage 7

7   a. Leaves cauline 8
    b. Leaves basal; one or two, flat on the ground 9

8   a. Petals fringed .... *Huttonaea*
    b. Petals not fringed .... *Centrostigma*

9   a. Flowers large, almost always solitary .... *Bartholina*
    b. Flowers small, 3 to many .... *Holothrix*

10 a. Lateral sepals each with a spur or pouch .... *Disperis*
    b. Lateral sepals without spur or pouch 11

11 a. Lip spurred 12
    b. Lip not spurred 20

12 a. Receptive surface of the stigmas on 2 projecting, club-shaped organs 13
    b. Receptive surface of the stigmas flat or concave, in a cavity below the rostellum 18

13 a. Flowers green, yellow, orange or green and white 14
    b. Flowers pink or white, often with darker markings 17

14 a. Petals divided 15
    b. Petals entire 16

15 a. Lateral sepals and lower petal lobes extensively fused to the stigmatic arms .... *Bonatea*
    b. Lateral sepals and lower petal lobes not fused to the stigmatic arms .... *Habenaria*

16 a. Flowers green, yellowish or green and white .... *Habenaria*
    b. Flowers orange, central rostellum lobe concave .... *Platycoryne*

17 a. Leaves numerous, in a basal rosette .... *Stenoglottis*
    b Leaves 1 or 2, near the base of the stem .... *Cynorkis*

18 a. Leaves scattered along the stem .... *Brachycorythis*
    b. Leaves basal or clustered in the lower stem portion 19

19 a. Petals as long as or longer than the sepals; inflorescence erect; leaves 1 or 2, flat on the ground .... *Holothrix*
    b. Petals shorter than the sepals; inflorescence normally nodding .... *Schizochilus*

20 a. Lip without an appendage 21
    b. Lip with a prominent appendage 31

21 a. Lip basally erect and fused to the column .... *Ceratandra*
    b. Lip basally not erect and not fused to the column 22

22 a. Stigmas 2, club-shaped .... *Stenoglottis*
    b. Stigma 1, not club-shaped 23

23 a. Leaves grass-like, linear or narrowly strap-shaped, basal 24
    b. Leaves not as above 25

24 a. Inflorescence apex nodding; sepals 2–4mm long .... *Schizochilus*
    b. Inflorescence erect; sepals mostly longer than 4mm .... *Disa*

25 a. Lip linear or narrowly strap-shaped, erect in front of the column and shorter than it .... *Brownleea*
    b. Lip spreading, as long as or longer than the column 26

26 a. Lip 3-lobed with prominent lobes 27
    b. Lip entire or with minute side lobes 28

27 a. Stem densely leafy .... *Brachycorythis*
    b. Stem only with sheaths .... *Schizochilus*

28 a. Lip fiddle-shaped, with large tooth-like apex .... *Disa*
    b. Lip linear to elliptic, apex not tooth-like 29

29 a. Leaf 1; petals about half as long as sepals; inflorescence slightly nodding .... *Dracomonticola*
    b. Leaves 2 to several; petals about as long as the sepals; inflorescence erect 30

30 a. Leaves 2, basal; median sepal unspurred; anther erect .... *Neobolusia*
    b. Leaves several; median sepal usually spurred; anther mostly reflexed to a horizontal or pendent position ...
    .... *Disa*

31 a. Lip anchor-shaped or broadly rhomboid; leaves cauline, linear and of about the same length 32
    b. Lip not as above; leaves various 33

32 a. Lip appendage 2-lobed; underground organ a cluster of thick roots with a furry sheath .... *Ceratandra*
    b. Lip appendage entire but apically bifid; plants with root tubers .... *Evotella*

33 a. Median sepal and petals forming a shallow, concave hood .... *Pterygodium*
    b. Median sepal and petals usually forming a deep concave hood; in one species shallow but then flowers
    non-resupinate, lip appendage with a prominent median keel on its front side and the plant from the
    Eastern Cape or southern Drakensberg .... *Corycium*

34 a. Plants terrestrial herbs, or rarely lianas rooting in the ground 35
    b. Plants epiphytic or lithophytic 48

35 a. Plants achlorophyllous, without green organs 36
    b. Plants not achlorophyllous 37

36 **a.** Flowers in upright position; flower stalk elongating to 200mm when fruiting .... *Didymoplexis*
   b. Flowers bell-shaped and drooping; flower stalk not elongating when fruiting .... *Gastrodia*

37 **a.** Lianas with leafless, green, succulent stems .... *Vanilla*
   b. Slender terrestrial herbs 38

38 **a.** Plants with a large single leaf appearing after flowering .... *Nervilia*
   b. Plants with several leaves that are always present at flowering time 39

39 **a.** Stems reed-like, 0.3–1m long; flowers white, with narrow sepals and petals over 45mm long ..... *Corymborkis*
   b. Stems not reed-like 40

40 **a.** Stems creeping in leaf litter and turning erect to flower 41
   b. Stems not as above 42

41 **a.** Lateral sepals almost free .... *Platylepis*
   b. Lateral sepals united for at least half of their length .... *Cheirostylis*

42 **a.** Column fused to the lip .... *Calanthe*
   b. Column not fused to the lip 43

43 **a.** Inflorescence terminal 44
   b. Inflorescence lateral 46

44 **a.** Median sepal and petals converging and forming a hood; flowers nearly hidden by bracts .... *Zeuxine*
   b. Median sepal and petals not forming a hood; bracts small 45

45 **a.** Lip spurless; plants pseudobulbous; leaves thin-textured .... *Liparis*
   b. Lip with a short spur; plants without pseudobulbs; leaves leathery .... *Acrolophia*

46 **a.** Pseudobulbs usually consisting of a single node; lip mostly 4-lobed .... *Oeceoclades*
   b. Pseudobulbs or underground corms consisting of more than one node; lip usually 3-lobed 47

47 **a.** Inflorescence usually lax; petals and sepals of different size, shape or colour .... *Eulophia*
   b. Inflorescence usually dense and often clustered near the apex; flowers bell-shaped, petals and sepals similar in size, shape and colour .... *Orthochilus*

48 **a.** Plants with fan of succulent and laterally compressed leaves; inflorescences terminal, with more than 50 small yellow flowers; from Limpopo .... *Oberonia*
   b. Plants usually with flat, keeled or cylindrical leaves or leafless; if rarely with a fan of succulent and laterally compressed leaves then inflorescences lateral and flowers less than 20; widespread 49

49 **a.** Pseudobulbs or swollen, cane-like stems present 50
   b. Pseudobulbs absent 53

50 **a.** Inflorescence lateral .... *Bulbophyllum*
   b. Inflorescence terminal 51

51 **a.** Flowers non-resupinate, small or medium-sized; median sepal less than 20mm long; roots all growing down .... *Polystachya*
   b. Flowers resupinate 52

52 **a.** Flowers large; median sepal more than 20mm long; at least some roots growing upwards to form a root-basket .... *Ansellia*
   b. Flowers small or medium-sized, median sepal less than 12mm long .... *Liparis*

53 a. Inflorescence terminal; growth sympodial; flowers non-resupinate .... *Polystachya*
   b. Inflorescences lateral, usually more than 1; growth monopodial 54

54 a. Plants with a fan of succulent, equitant and laterally compressed leaves .... *Bolusiella*
   b. Plants not with fan of succulent leaves; leaves flat, keeled, cylindrical or absent 55

55 a. Flowers yellow with purple markings; spur pouch-shaped .... *Acampe*
   b. Flowers uniformly white, cream-coloured or pale green, sometimes with darker tinges on tip of tepals and
      spur; spur various 56

56 a. Plants leafless 57
   b. Plants with green leaves 59

57 a. Spur 1–6mm long .... *Microcoelia*
   b. Spur 17–45mm long 58

58 a. Anther-cap bright green; perianth bell-shaped with broadly rounded lobes .... *Margelliantha*
   b. Anther-cap yellow, brown or dull green; perianth lobes narrow and pointed .... *Mystacidium*

59 a. Lip wide at the base and encircling the column .... *Angraecum*
   b. Lip arising from below the column, and normally not encircling it 60

60 a. Flowers 1 or rarely 2 per inflorescence; rostellum short and notched .... *Jumellea*
   b. Flowers normally 3 or more per inflorescence; rostellum mostly beaked 61

61 a. Flowers white; apical half of lip not or only shallowly 3-lobed 62
   b. Flowers green, cream, pale brown or yellow; apical half of lip various 68

62 a. Anther-cap bright green; perianth bell-shaped 63
   b. Anther-cap yellow, brown or dull green; perianth not bell-shaped 64

63 a. Leaves 25–65mm long .... *Margelliantha*
   b. Leaves 100–150mm long .... *Diaphananthe*

64 a. Sepals and petals narrow and pointed, similar to the lip .... *Cyrtorchis*
   b. Sepals and petals not as above 65

65 a. Leaves folded near the base but flat and usually broadening towards the apex; stipe 1 .... *Aerangis*
   b. Leaves V-shaped all along their length; stipes 1 or 2 66

66 a. Stipe 1, Y-shaped, divided for nearly half the length; spur usually less than 45mm long .... *Ypsilopus*
   b. Stipes 2, free to near the base; spur 8–90mm long 67

67 a. Mouth of the spur narrow; the two pollinia attached to the same viscidium .... *Rangaeris*
   b. Mouth of the spur wide, each pollinium attached to an individual viscidium .... *Mystacidium*

68 a. Lip apically deeply 3-lobed, its side lobes often fringed; stipe 1, viscidium 1 .... *Tridactyle*
   b. Lip apically not deeply 3-lobed; stipes 2, viscidia 1 or 2 69

69 a. Lip with a conspicuous tooth in the mouth of the spur; the two pollinia attached to the same viscidium ...
      .... *Diaphananthe*
   b. Lip without a tooth in the mouth of the spur; each pollinium attached to an individual viscidum 70

70 a. Flowers not transparent; lip narrowly triangular, acute; spur more than 10mm long .... *Mystacidium*
   b. Flowers semi-transparent; lip fan-shaped; spur less than 6mm long .... *Rhipidoglossum*

GLOSSARY

achlorophyllous: pertaining to a plant that lacks chlorophyll and obtains all nutrition from organic matter through a symbiotic, non-pathogenic association between certain fungi and the roots of plants. These plants are often referred to as saprophytes although scientifically this is not quite accurate.

apiculate: ending in an abrupt, sharp point

apiculus: short, sharp but not stiff point

auricle: small, ear-shaped lobe

bifid: divided or lobed into two

blade: flat, wide portion of a leaf or flower part

bract: small leaf on the inflorescence at the base of a flower or flower stalk

callus (pl. **calli**): fleshy outgrowth or tissue thickening, particularly on the lip

cauline: scattered along the stem

column: central structure consisting of the fused style and stamens and thus containing the stigma and pollen; also called gynostemium

column foot: extension at the base of the column to which the lip is attached

corm: thickened, bulb-like underground stem

crisped: with an irregular, wavy margin

decurved: curved downward and outward

deflexed: bent abruptly downward

disc: flat surface of the lip

disjunct: with clearly separated distribution areas

distal: furthest from the point of origin or attachment; towards the apex

endemic: native and restricted to a particular, defined area

entire: with a smooth, even margin; not divided

epiphyte: a plant growing on, or attached to, another plant without deriving water or nourishment from it

equitant: with the base of one leaf clasping the base of the opposite leaf above, which in turn clasps the one above that

exserted: projecting beyond the surrounding parts

flexuose: wavy in a zigzag manner

galea: helmet-like structure derived from one of the perianth lobes

habit: the general appearance or characteristic growth form

hysteranthous (of leaves): developing after flowering

limb: stalk-like portion below the blade; lip and petals may have a pronounced limb

lip: median, often modified petal; may be different in colour, shape or size compared to the lateral petals; also called labellum

lip flap: extension of the lip

lithophyte: a plant growing on rocks

mentum: chin-like extension at the base of a flower consisting of the column foot, lip and lateral sepals variously joined

monopodial: of a plant habit with unlimited growth of the main axis

non-resupinate: not upside down; with the lip uppermost

oblong: elongate and rather narrow; 2–4 times longer than broad, with nearly parallel sides

ovoid (3-dimensional): egg-shaped

pantropical: occurring in all tropical regions of the world

papilla (pl. **papillae**): small, fleshy outgrowth

papillate: bearing papillae

papillose: bearing many papillae

perianth: the floral parts around the column, i.e. the sepals, petals and lip

pollinium (pl. **pollinia**): a mass of united pollen grains

pseudobulb: thickened part of stem resembling a bulb

rachis: the flower-bearing part of the inflorescence, above the peduncle (stalk)

radical (of leaves): arising so close to the base of the stem as to appear to come directly from the roots

recurved: gradually bent or curved backward or downward

reflexed: bent sharply backward or downward

resupinate: upside down, with the lip lowermost, as in most orchids

rhizome: underground stem, distinguished from roots by its nodes, buds or scale-like leaves

rhomboid: more or less diamond- or lozenge-shaped; in the shape of an equilateral parallelogram

rostellum: shelf- or beak-like sterile projection on the column that separates the stigmatic surface from the anther

rostellum arms: arm-like outgrowths of the rostellum

rudimentary: small and non-functional; poorly developed

sessile: directly attached; without a stalk

stigmatic arms: protuberances on which the stigmas are located

stipe: pollinium stalk

suffused: tinted or tinged with colour; spread throughout with colour

sympodial: of a plant habit in which each shoot has limited growth and new shoots arise from the base of the old one

tepal: segment of the perianth, i.e. petal or sepal

theca (pl. **thecae**): pollen sac of the anther

ventral: on the lower surface

viscidium (pl. **viscidia**): sticky gland to which the stipe and pollinia are attached

FURTHER READING

Ball, J.S. 1978. *Southern African Epiphytic Orchids.* Conservation Press, Johannesburg.

Bolus, H. 1893–1913. *Orchids of South Africa – Icones orchidearum austroafricanarum extra-tropicarum; or figures, with descriptions, of extra-tropical South African Orchids.* Vols. 1, 2 & 3. William Wesley & Son, London.

Harrison, E.R. 1972. *Epiphytic Orchids of Southern Africa.* Natal Wildlife Protection and Conservation Society of South Africa, Durban.

Liltved, W.R. & Johnson, S.D. 2012. *The Cape Orchids: A Regional Monograph of the Orchids of the Cape Floristic Region.* Sandstone Editions, Noordhoek.

Linder, H.P. & Kurzweil, H. 1999. *Orchids of Southern Africa.* Balkema, Rotterdam.

Manning, J. & Paterson-Jones, C. 2004. *Southern African Wild Flowers: Jewels of the Veld.* Struik Nature, Cape Town.

McMurtry, D., Grobler, L., Grobler, J. & Burns, S. 2008. *Field Guide to the Orchids of Northern South Africa and Swaziland.* Umdaus Press, Hatfield.

Mucina, L. & Rutherford, M.C. (eds) 2006. *The Vegetation of South Africa, Lesotho and Swaziland.* Strelitzia 19. South African National Biodiversity Institute, Pretoria.

Pooley, E. 1998. *A Field Guide to Wildflowers of KwaZulu-Natal and the Eastern Region.* Natal Flora Publications Trust, Durban.

Schelpe, E.A.C.L.E. 1966. *An Introduction to the South African Orchids.* MacDonald & Co., London.

Stewart, J. & Hennessy, E.F. 1981. *Orchids of Africa: A Select Review.* Macmillan, Johannesburg.

Stewart, J., Linder, H.P., Schelpe, E.A. & Hall, A.V. 1982. *Wild Orchids of Southern Africa.* Macmillan, Johannesburg.

Web-based resources

BRAHMS. Botanical Research and Herbarium Management System
http://herbaria.plants.ox.ac.uk/bol/

Flora of Zimbabwe
http://www.zimbabweflora.co.zw

iSpot **http://www.ispotnature.org/communities/southern-africa**

PlantZAfrica.com **http://www.plantzafrica.com**

Red List of South African Plants
http://redlist.sanbi.org/

SABIF (South African Biodiversity Information Facility) **http://www.sabif.ac.za**

SIBIS (South African National Biodiversity Institute Integrated Biodiversity Information System)
http://sibis.sanbi.org

Swaziland's Flora Database
http://www.sntc.org.sz/flora/

Swiss Orchid Foundation at the Herbarium Janny Renz **https://orchid.unibas.ch/**

World Checklist of Selected Plant Families
http://apps.kew.org/wcsp/home.do

Disa maculata

INDEX

Accepted names are in **bold italic**. Synonyms are in *italic*.